A HISTORY OF

# GREEN RIDGE STATE FOREST

A HISTORY OF

# GREEN RIDGE STATE FOREST

*Champ Zumbrun*

THE
History
PRESS

Published by The History Press
Charleston, SC 29403
www.historypress.net

*Front cover*: Photograph by Tom Darden, Maryland governor's office.
*Back cover, clockwise from top left*: Photograph by Tom Darden, Maryland governor's
office; courtesy of Tom Hynson; American Forests, 734 15th Street NW, suite 800,
Washington, D.C. 20005. www.americanforests.org.

First published 2010

ISBN 9781540234940

Library of Congress Cataloging-in-Publication Data
Zumbrun, Francis.
A history of Green Ridge State Forest / Francis Zumbrun.
p. cm.
ISBN 978-1-59629-902-3
1. Green Ridge State Forest (Md.)--History. 2. Allegany County (Md.)--History.
3. Allegany County (Md.)--Biography. 4. Allegany County (Md.)--Environmental
conditions. 5. Mountain life--Maryland--Allegany County--History. 6. Outdoor
life--Maryland--Allegany County--History. 7. Forests and forestry--Maryland--
Allegany County--History. I. Title.
F187.A4Z86 2010
975.2'94--dc22
2010014772

*"Remember the days of old,*
*Consider the years of all generations.*
*Ask your father, and he will inform you,*
*Your elders, and they will tell you."*
*—Deuteronomy 32:7*

# Contents

# Contents

# PREFACE AND ACKNOWLEDGEMENTS

*If I have seen further than other men, it is by standing on the shoulders of giants.*
*—Sir Isaac Newton*

There are so many people to thank for making this book possible. First, I would like to thank my wife, Cindy, for all her suggestions and support that allowed me the time to write this book. In addition, I would like to thank my sons, Jeffrey and Ryan, for their support over the years.

Many of those people I thank I do not know personally, as they came before me, especially those individuals who served in the Maryland Forest Service in the 1950s and before. These pioneers planted the seeds of forest conservation in Maryland, established, and made credible the new science of forestry and, because of their hard work in forest conservation, ultimately restored Maryland's devastated forests back to health.

I have met and have had the privilege to know many of the individuals who worked with the Maryland Forest Service since the 1960s. They, too, are pioneers in their own right as they all have worked in the first century in which scientific forestry was practiced in America. The agency often changed as new scientific breakthroughs came in the field of forestry and related conservation

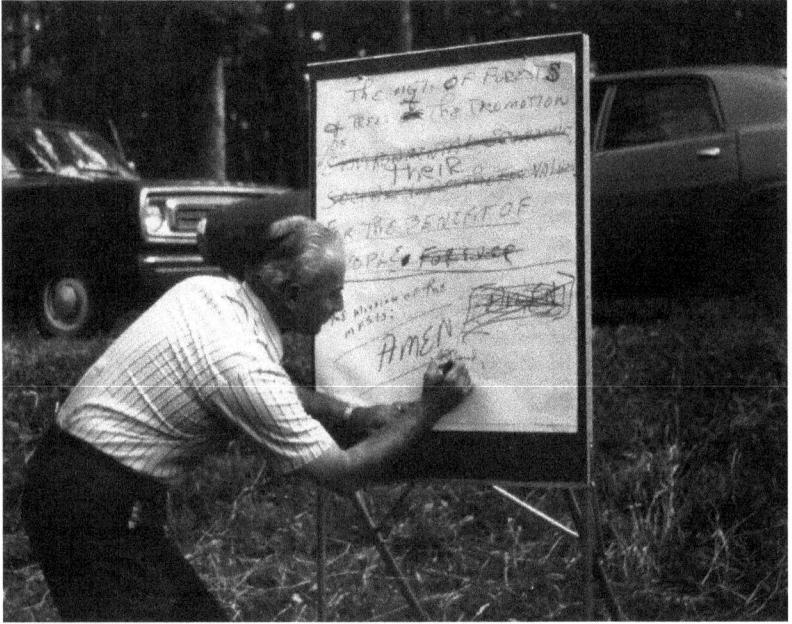

Adna "Pete" Bond, state forester (1968–77), signing off on an early Maryland Forest Service mission statement, circa 1970s. *Courtesy Jenny Bond.*

disciplines. They had to be resilient, persistent and creative as they worked hard with limited budgets and resources to carry out the Maryland Forest Service Mission that "restores, manages, and protects Maryland's trees, forests, and forested ecosystems to sustain our natural resources and connect people to the land."

Although it is impossible to mention everyone, I would like to thank all those who worked at Green Ridge State Forest (GRSF) over the years. GRSF serves as a microcosm for forest conservation in Maryland. Foresters divide the history of forest conservation into five broad periods: the Age of Forest Exploitation (1860–1906), the Custodial Period (1906–42), the Sustained Yield Management Period (1942–70), the Multiple-Use Management Period (1970–90) and the Sustainable and Forest Health Period (1990–present). The individuals are mentioned generally in the context of the periods they worked. All the individuals contributed in some way to this book. I asked questions about the days of old, and they answered me, through writings and documents they left behind or through oral history interviews and casual conversations with friends.

The following discussion uses the proverbial goose and golden egg analogy to illustrate changes that occurred over the past century. The goose represents the forest, and the egg represents forest values derived from forest resource management (timber production, recreation opportunities and water and wildlife conservation. Further benefits also include enhancements for air quality and scenic quality in addition to economic stimulants that are both direct and indirect).

## THE AGE OF FOREST EXPLOITATION (1860–1906)

This period preceded the Maryland 1906 Forestry Conservation Act. Beginning about the 1860s, the volume of timber cutting greatly exceeded growth. By the 1900s, only 20 percent of mature forest cover existed east of the Mississippi River in the United States. The public feared a timber shortage. Large, uncontrolled forest fires usually followed the cutting. The forests of Maryland consisted primarily of stumps, seedlings and saplings. Wildfires were not controlled and burned over large areas. Farmers allowed domestic livestock to graze in the woodlands, destroying young trees. The animals also compacted the soil, bruised seedlings and saplings and destroyed wildlife habitat. It was very unusual to see large trees. In Fred W. Besley's early glass lantern slides, many photographs show people posing next to large trees. To see a large tree back then was equivalent to seeing a bald eagle today—it was very special. Because of his interest in large trees, Besley started a Maryland Champion Big Tree contest that today is a national program. Forest health was poor. The goose was sick, producing few eggs.

## THE CUSTODIAL PERIOD (1906–42)

In general, the Custodial Period covers the entire administration of Fred W. Besley as Maryland's first state forester. Besley's mission was to restore, conserve and protect the forests and establish state parks and forest recreation areas. Much of his efforts involved protecting the forests from wildfire, overseeing and improving timber-harvesting practices, controlling livestock grazing in forest woodlands and beautifying roadsides with trees. Besley's administration surveyed, inventoried and mapped the entire forests of Maryland for the first time. He established Maryland's first state tree nurseries. Conservationists from all occupations planted millions of trees grown at the state nursery on Maryland's abandoned agricultural fields and devastated forestland. Besley oversaw the construction of fire lookout towers across the state to monitor and respond to outbreaks of wildfires. Officials trained forest wardens in forest fire prevention, suppression and public awareness. During the Great Depression, President Roosevelt's Civilian Conservation Corps (CCC) came into existence (1933–42). Besley established three CCC camps at GRSF. The State of Maryland actively acquired cutover and abused land for public use, building up the state park and state forest system. The CCC constructed roads, fought fires, erected fire lookout towers and planted millions of trees that significantly improved the overall health of public lands. During this period, Maryland's forest began to recover from past abuses that had occurred during the period of exploitation. The state established GRSF in 1931, and the Town Hill Lookout fire tower was built in the same year. GRSF did not require a full-time state forester at first, as the forest only consisted of about two thousand acres. The first full-time forester was hired in 1941, when the forest grew to more than ten thousand acres. The health of the goose slowly improved, and it began producing a small number of eggs on a regular basis.

GRSF Staff: Herb Robertson (GRSF's first state forester, 1941–42), H.B. Buckingham (district forester of western Maryland), Frank Davis (forest warden), Urner Wigfield (forest warden), Eugene P. Sipes (first resident forest warden and CCC foreman), Clem Wigfield (fire tower watchman), George Wigfield (forest warden) and Henry D. Schaidt (supervisor). At the Green Ridge CCC Camp (S-53) were the following employees: David W. Sowers (project superintendent), H.L. Stuart (engineer), H.K. Cheney (mechanic), Mr. Wigfield (blacksmith), C.W. Shagel (clerk), Cecil Ward (foreman), C.V. Crocks (foreman), R.H.J. Wempe (foreman) and Fred W. Besley (state forester, 1906–42).

## Sustained Yield Management Period (1942–70)

This period included events of World War II and the arrival of the baby boomers. Forest harvests equaled annual growth. A great demand for wood products occurred during this intense period of economic growth. The population of the United States quickly grew. The rural society was becoming urban, and more technical forest fire control methods were advanced. In addition, cold war civil defense procedures were intensified within the forest fire control staff. The foresters made management decisions independently without public participation. Foresters researched, surveyed and made accurate hand-drawn maps showing the location of the ten-acre tracts of land with unknown owners known as the "the Lost Souls," formerly part of the Mertens Green Ridge Valley Orchards. In 1969, the forest manager oversaw the first large timber harvest since 1931, when the state forest was established. The forest manager called this harvest, more than seventy acres in size, a "clean-cut." This timber harvest was located at the crossroads of Oldtown Road and Mertens Avenue. The goose's health had returned, and public recreation demands increased. The trees were now merchantable, and forest management focused its attention on one golden egg—timber.

GRSF staff and regional staff support: Harry Hartman (resident forester, 1949–66), Ray Johnson (forest manager, 1966–71), Brooke Botkin (forest warden and fire control), Floyd Custer (superintendent, 1956–76), Clint Irwin (district forester for western region), William H. Johnson (district forester for western region, 1943–60), R. Thomas Thayer Jr. (district Forester for Western region), Ralph Peace (district park superintendent for western region), Joseph F. Kaylor (state forester, 1942–47) and Henry C. Buckingham (state forester, 1947–68).

## THE MULTIPLE-USE MANAGEMENT PERIOD (1970–90)

This period marks the beginning of an interdisciplinary approach to forest resource management. A variety of natural resource disciplines began providing the forest manager with management recommendations. The environmental era began with the beauty and preservation of Maryland's natural resources being the primary concern. In 1969, Program Open Space was established and provided opportunities and funding to purchase additional forested public land. In 1971, wildland laws were enacted. Savage River State Forest was the first public land with a wildland designation. Green Ridge State Forest followed not long after with the designation of the Potomac Bend Wildlands. Foresters oversaw even-aged managed harvests and white pine release harvests. Staff established the GRSF hiking trail, constructed one hundred campsites and built the GRSF shooting range. Forest management considered social values as

Ray Johnson (left) and Offutt Johnson (right) helped pioneer the Program Open Space program that built up public lands in Maryland, circa 1970s. *Courtesy Ray Johnson.*

well as economic values. The management focus evolved from one egg (timber) to a basket of eggs (timber production, wildlife, water and soils conservation, recreation opportunities, further air quality enhancement, improved scenic quality and economic stimulation).

GRSF staff and regional staff support: John Mash (GRSF forest manager, 1971–88), George Gilmore (regional forest manager), Harland Uphole (regional fire manager), Walter "Pete" Glass (regional fire manager), Hobart Bennett (mechanic), Marvin Broadwater (mechanic), Paul Williams (ranger), Marshall Fletcher (ranger), Carl "Bill" Shipley (chief of maintenance), Greg Sherwood (forest technician), Bill Slider (fire control), John McCusker (labor foreman), Perry Edminston (administrator), Harold McDonald Sr. (regional maintenance supervisor), Harold McDonald Jr. (forest ranger), Wayne Rice (fire control), Gloria Dilley (office administrator), Anne Drumme (western regional administrator), Offutt Johnson (Program Open Space), Carol Lease, Adna R. Bond (state forester, 1968–77), Donald E. Maclaughlan (state forester, 1978–79), Tunis Lyon (state forester, 1979–83) and James B. Roberts (state forester, 1983–91).

## Sustainable and Forest Health Period (1990–present)

Today, forest management closely considers ecological processes, biodiversity and forest health. Outputs like recreation opportunities and timber production, for example, are a byproduct of forest health and sustainable forest use. Annual growth in the forest now greatly exceeds harvests (note: in Maryland's state forests, growth is more than four times the harvest). The majority of Maryland's population is urban. Public participation now consists of interdisciplinary teams (Department of Natural Resource professionals), advisory committees (composed of citizens from a variety of backgrounds and interests) and the public, who attend informational meetings held in the community. The forest manager on state forests writes long-range management plans with input from the public. Today, forest managers examine the forest on a much broader scale, expanding their view beyond state forest property to the surrounding area within a watershed. At the same time, foresters are using a new generation of tools to get the job done, including satellite imagery, computers, geographic information systems (GIS) and global positioning systems (GPS). The focus is now on the goose (forests) and its health. The eggs (forest products) are secondary to the health of the forest. In conclusion, a healthy forest makes all the other aforementioned benefits possible.

GRSF staff and regional staff support: Francis Zumbrun (GRSF forest manager, 1988–2009), John Mash (regional forest supervisor), Bill Cihlar (Rocky Gap State Park manager), Chris Anderson (regional state park manager), Bob Webster (western regional forest manager), Barbara Stevens (regional administrator), Ric Lillard (western regional fire manager), Bernie Zlomek (Allegany Project forester), Randy Kamp (forest ranger), Eric Warnick (forest ranger), Zeke Seabright (forest ranger) and Phil Pannil (Forest Pesticide expert). GRSF employees during this period included John Denning (assistant forest manager), Mike Schofield (assistant forest manager), Mark Beals (assistant forest manager), Jesse Morgan (forester), Donald Smith (maintenance chief), Robyn Roland (office administrator), Dennis Yoder (chief of maintenance), Tom Grieves (mechanic), Harry Cage (Maryland ranger), John Amann (Maryland ranger), Scott House (Maryland ranger), Jeffrey Ruark (Maryland ranger), Mike Deckelbaum (Maryland ranger), Curt Dieterlie (Maryland ranger), Rick Lewis (Maryland ranger), Tom Acton (Maryland ranger), Jeff Herndon (Maryland ranger), Shelley Miller (naturalist), Alicia Norris (naturalist), Roy Musselwhite (forest technician), John Braskey (Program Open Space), George Forlifer (property management manager) and members of the GRSF DNR Interdisciplinary Team, Deidra Ritchie (Program Open Space), Tom Grieves Jr. (conservation aide), Eric Schwaab (DNR deputy secretary), Joe Gill (DNR deputy secretary), John Griffin (DNR secretary), Dave Moore, John Moore, Harvey Bryant, Harry Tusing, Rick Barton (state park superintendent), Nita Settina (state park superintendent), John W. Riley (state forester. 1991–95), James E. Mallow (state forester, 1995–2001), Stephen W. Koehn (state forester, 2001–present), and Bill Buck (state forest environmental advisor).

Other individuals I want to thank outside the Maryland Forest Service who made this book possible include: the Cresap Society and descendants of Thomas Cresap, especially Nina Higgins, Karen Cresap, Michael Cresap, Paul Vermillion and Jeanne Bryant; Bob Bantz and Homer Hoover of the Western Maryland Archaeology Society; Mary Rotz; Helen Besley Overington; Kirk Rodgers; Peggy and Donald Weller; John Overington and the descendants of the Besley and Rodgers Corporation; Tom Hynson, a descendant of Fred Mertens, for graciously giving permission to use his Mertens family archival photographs; Debbie Myers, Jan Alderton and Mike Sawyers of the *Cumberland Times-News* staff; Jim Mullan and Tom Mathews, DNR Wildlife Division; Gwen Brewer and Ed Thompson, DNR Heritage; Patty Manown-Mash, Offutt Johnson, Melissa McCormick, Jack Perdue, Ross Kimmel, Robb Bailey, Linda Wiley, Ann Wheeler, Christina Holden, Roberta Dorsch, Chuck and Erin Thomas and members of the Committee for Maryland Conservation History; Tom Darden and staff with the Maryland Governor's Photography Office; Alexander Clark and Harry Kaylor, DNR Inventory crew; Jane R. Foster and Dr. Phil Townsend from the Appalachian Laboratory in western Maryland; Bill Schoenadel, Denise and Lew Langham and the Friends of GRSF; Dale Sipes, charter chair, Brad Metzger and members of the GRSF Advisory Committee; Herb Schwartz, Dennis Tipton and the Volunteer team Inc.; Peter and Karen Miller of the Maryland Forests Association; Steve Resh and the Maryland-Delaware Chapter of the Society of American Foresters; Curtis Clark and Brothers; Cessna Brothers; Paul E. Smith, John Whorton and GRSF Forest Product operators; Mike Lewis, Gary Carpenter, Lonnie Lewis and Maryland Department of Juvenile Services staff; Jeff Horan, chief of Water Resources; Harry Kahler and Alexander Clark (forest inventories); Robb Schoeberlein and Richard Richardson (Maryland State Archives); Eve Higman, Frenis and Mabel Hoffman of the Living History Foundation; and Al Feldstein, western Maryland historian.

Finally, I would like to thank the staff at The History Press, especially Hannah Cassilly, commissioning editor, and Hilary McCullough, editorial department manager, for their patience in guiding me step by step through the process of publishing my first book.

Introduction

# GREEN RIDGE: A PLACE OF STATE AND NATIONAL SIGNIFICANCE

*[Allegany County is a] region steeped in a rich history of human occupancy. Not only does the county bear the nation's first superhighway—The Potomac River—it also holds mounds of archaeological resources. It's a unique place—a place you all love. People have loved this place for a thousand years.*
*—Charles Hall, state archaeologist for the Maryland Historic Trust*

The staff at Green Ridge State Forest (GRSF) has heard many public comments over the years about what the state forest meant to them. One finds out quickly that Green Ridge is many things to many people, judging by a sampling of their remarks: "Due to its size, Green Ridge is surely one of the of the chief cornerstones of Maryland's public land systems." "Green Ridge State Forest is one of Maryland's last great forested frontiers." "Green Ridge State Forest is the people's forest." "It is Baltimore's playground." "Green Ridge is the land of the living." "It is an enchanted, awesome place." "Green Ridge is a place of national and state significance, one of Maryland's shining crown jewels." "Green Ridge is no walk in the park." "Green Ridge is one of Maryland's treasured landscapes, one of the last best places."

The story of how Green Ridge State Forest became a state forest is just as remarkable. The people who walked its landscape over the centuries to the present time are just as extraordinary. This story focuses on a sampling of several pioneers who graced this land:

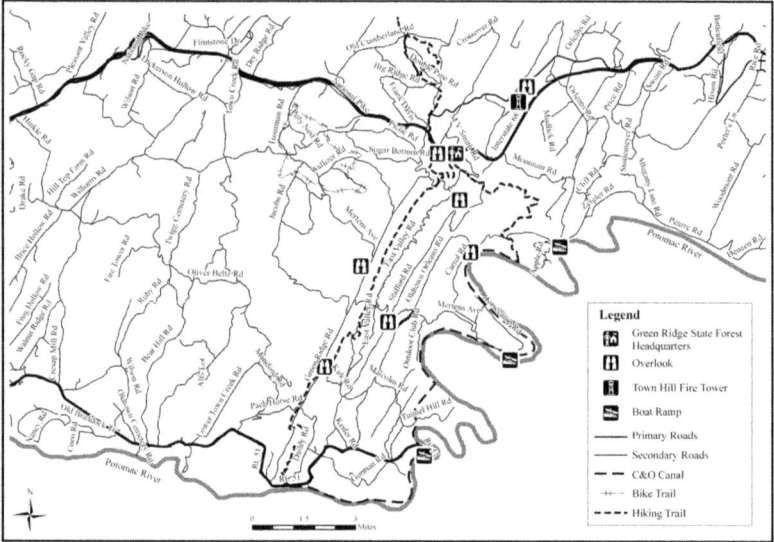

Green Ridge State Forest map. *Courtesy Jesse Morgan and Nicole Sisler.*

Thomas Cresap, the frontiersman; the Carroll family and Mertens family, business entrepreneurs; Fred Besley, Maryland's first state forester; and Jane R. Foster, forest researcher. They are all pioneers and trailblazers in their own right. Through their stories, we will also learn a little about the land that came to be Green Ridge State Forest. However, before we discuss these personalities, a brief overview of the state forest is in order.

Green Ridge State Forest is located in eastern Allegany County, Maryland. It is the only state forest located in the Ridge and Valley physiographic province. Green Ridge receives the least amount of rainfall in Maryland, averaging thirty-six inches annually. Consisting of about 43,560 acres, it accounts for about 29 percent of the state forest system and about 12 percent of all DNR land in Maryland.

The general geographic boundaries of Green Ridge are Town Creek to the west and Sideling Hill Creek to the east. The northern boundary extends to the Mason-Dixon line. The southeastern boundary parallels the Potomac River. Elevations range from five hundred feet above sea level on the Potomac River to two thousand feet on Town Hill. Two major highways, Interstate 68 and MD Route 51, traverse the forest in an east–west direction.

Native Americans have occupied the region for more than ten thousand years. Thomas Cresap (1694–1787), Maryland's great pioneer, patriot and pathfinder, was the first European American to settle permanently in the area, at Oldtown, Maryland, about 1740.

In the early 1800s, Richard Caton (namesake for Catonsville, Maryland) and William Carroll formed a partnership and owned much of the land that is Green Ridge State Forest today. Richard Caton was the son-in-law to Charles Carroll of Carrollton, a signer of the Declaration of Independence. William Carroll was the grandson of Daniel Carroll of Rock Creek, who was a framer of the United States Constitution.

The Carroll family patented most of the vacant lands in the Green Ridge area during the 1820–40 period for inclusion into various timber and mining interests, primarily the Town Hill Mining, Manufacturing and Timber Company. The estate of Charles Carroll of Carrollton financed this business venture. A crumbling stone structure known as the Carroll Chimney, part of a steam-powered sawmill built in 1836, is the only surviving structure from this period. In the 1880–1912 era, most of the remaining original virgin forest disappeared during a period of overcutting and abuse, the neglect resulting in numerous wildfires.

The Mertens family of Cumberland purchased the land from the Carroll heirs in the late 1800s, keeping the large tract of land under one ownership. The Mertenses came from a German shipbuilding family, arriving in Cumberland, Maryland, in 1851. The canalboats the Mertenses built were some of the first to operate on the Chesapeake and Ohio Canal. During the early 1900s, the Mertenses converted the forest to apple orchards and promoted it as the "largest apple orchard in the Universe." The Mertenses subdivided the orchard into ten-acre parcels and sold them to individuals as investment properties. Five acres in each parcel were cleared, burned and planted into apple trees. In the remaining five acres, loggers cut the best quality trees, leaving the poorer quality trees standing. The Mertenses sold ten-acre tracts to more than three thousand people before they went bankrupt in 1918.

Under the direction of Fred W. Besley (1872–1960), Maryland's pioneering first state forester, the State Department of Forestry began acquiring the interests of the corporation. In 1931, Green Ridge State Forest was established. Some of the first major forest management activities at Green Ridge State Forest were performed by the Civilian Conservation Corps (CCC) from 1933 to 1942, during which time three CCC camps were located at GRSF. The CCC worked on forest fire control activities, built roads and trails, enhanced recreation infrastructure and managed the forest for its future timber and wildlife potential.

During World War II, the CCC camp located at Fifteen Mile Creek housed German prisoners of war, who worked on nearby fruit orchards and cut pulpwood in the forest. As the forest matured, GRSF became more and more popular with outdoor enthusiasts, especially hunters. It also contributed increasingly to the local wood products industry and the local and regional economy.

The Maryland Department of Juvenile Services established the Boys Forestry Camp in 1955 at the same location of the former Green Ridge CCC camp along Fifteen Mile Creek and, in the 1970s, the Maple Run Boys Forestry Camp off Jacobs Road. Over the years, the youth at these camps assisted the Maryland Forest Service and Department of Natural Resources on various forest, recreation and wildlife conservation projects at Green Ridge.

Since the establishment of the state forest in 1931, scientists have accomplished incredible groundbreaking research regarding the forest, its wildlife and the environment that have contributed nationally to the general knowledge of the public.

Today, GRSF is an approximately 110-year-old, even-aged, mixed-oak hickory forest. The oak consists of a variety of species characteristic of dry, upland areas that include black oak, white oak, red oak, scarlet oak and chestnut oak. Five native pines grow at Green Ridge State Forest: white pine, Virginia pine, pitch pine, table-mountain pine and short-leaf pine. Flowering dogwood, redbud and serviceberry are common understory trees.

Upland animals found in abundant numbers in the forest are white-tailed deer, fox and gray squirrels, raccoons, red foxes and cottontail rabbits. Other animals include muskrats, beavers, minks, chipmunks, mice, flying squirrels, weasels, skunks, opossums, bobcats and black bears. Wild turkey and ruffed grouse are popular game birds in Green Ridge. Other birds include the pileated woodpecker, red-tailed and broad-winged hawk and the barred owl. A wide variety of Neotropical migrants and songbirds also occur in the forest. Wildflowers such as May apple, coltsfoot, spring beauty, trillium, bloodroot and spiderwort flourish at Green Ridge.

Green Ridge State Forest in eastern Allegany County is one of Maryland's last remaining forested frontiers. Encompassing 43,560 acres, it is the state's largest contiguous block of forestland within the Chesapeake Bay watershed. Landmarks and personalities associated with GRSF make the forest a place not only of statewide significance but of national significance as well.

The National Road—"The Road that Built the Nation"—passes through GRSF and is one of the best-known historic and scenic cultural roadways in the United States. The Maryland National Road Association recognizes the seven-mile section between Belle Grove and Fifteen Mile Creek as the most picturesque section along the Historic National Road Scenic byway in the state. Early twentieth-century postcards called the Town Hill section of the National Road "the Beauty Spot of Maryland," complete with breathtaking scenery described as "wild and beautiful beyond anticipation."

Looking east from Town Hill, just north of the Sideling Hill I-68 road-cut, is the Mason-Dixon line, the most famous border in America. Extending over a multistate area, the famous boundary line separated the North (the free states) from the South (the slave states) prior to and during the American Civil War. The Mason-Dixon line in eastern Allegany County generally serves as the northern border of GRSF.

Looking southeast from Town Hill, one can see the Potomac Gap where the Potomac River, called "the nation's river," cuts through Sideling Hill as it continues its southeastward flow to Washington, D.C. One of most beautiful and bountiful rivers on the East Coast, the Potomac flows along the GRSF Potomac Bends Wildland and makes up about thirty miles of the forest's eastern and southern boundaries.

Paralleling the Potomac River and Green Ridge State Forest is the 184.5-mile Chesapeake and Ohio (C&O) Canal National Historical Park, stretching from Cumberland to Georgetown. The GRSF section of the canal includes the Paw Paw Tunnel, promoted by its builders in the mid-nineteenth century as a "wonder of the world." Roughly 3,118 feet long, it is the largest fabricated structure on the entire canal. Once inside, a hiker is plunged into darkness, able to see only a small circle of light at the tunnel's far end (I suppose this experience is as close as one can get to a near-death experience without being near death).

The Potomac Heritage National Scenic Trail System includes the C&O Canal National Historical Park as well as the Green Ridge State Forest Hiking Trail. Former U.S. Secretary of the Interior Gail Norton designated the GRSF trail as a National Recreation Trail in June 2005, and just recently, in May 2006, it became part of the Great Eastern Trail that stretches more than 1,600 miles and connects more than 10,000 miles of foot trails.

A place of national significance should have a national figure associated with it, and Green Ridge has just that: George Washington. The father of our country visited this region during every aspect of his colorful career: as a surveyor, military officer, landowner and later as the first president of the United States. Point Lookout offers a strikingly powerful promontory view of the surrounding landscape on the eastern side of GRSF. From here, one can view two-hundred-plus acres of land that George Washington once owned. The ancient Potomac River's serpentine waterway seen from Point Lookout in 1840 was known as "General Washington's Horseshoe Bend."

When Washington was in the area, he often stayed overnight with Colonel Thomas Cresap at Oldtown, the county's first frontier settlement. Like Daniel Boone in Kentucky, Thomas Cresap was Maryland's great pioneer, pathfinder and patriot. In addition to helping blaze the National Road in 1753, Cresap forged the Oldtown Road through Green Ridge in 1758, connecting Fort Frederick to Fort Cumberland during the French and Indian War.

The population of Maryland has almost tripled since 1931, the year Besley established the state forest, from about two million to

Point Lookout. Photograph taken by Fred W. Besley, state forester, circa 1920s. Note the water in the canal. *Courtesy Jenny Bond.*

six million people. Millions of people in surrounding states are now within a couple hours' drive of Green Ridge State Forest. As a result, Green Ridge has become an important place where people can escape urban life to reconnect with nature to relax, revitalize and rejuvenate themselves. Hunting, mountain biking, camping, canoeing, bird-watching and sightseeing are just some of the popular recreational activities that occur in the forest. Great accolades have been received from visitors, such as the following: "Green Ridge State Forest is one of the best backpacking outings in Maryland at any time of the year," Leonard M. Atkins, author of *50 Hikes in Maryland*; "Green Ridge is Maryland's mountain bike Mecca," Bryan Heselbach, mountain bike event coordinator; "Camping in Green Ridge is almost too good to be true. It also happens to be, for what it's worth, my favorite camping destination in Maryland. This is because it's only two hours from my home in Baltimore, lacks the massive crowds heading east to the ocean and the better-known recreation areas to the west and allows for truly spectacular, away-from-it-all camping," Evan L. Balkan, author of *The Best Tent Camping in Maryland*.

Recreation at Green Ridge State Forest also plays a major role in western Maryland's economic growth and tourism industry. Annual expenditures by outdoor enthusiasts in Maryland for fishing, hunting and wildlife watching totaled $1.4 billion in 2001. A total of 645,000 anglers went fishing in Maryland an average thirteen times in 2008 and spent $568 million; 161,000 hunters went out an average of fourteen times in Maryland and spent $210 million; 1.5 million people went wildlife watching in Maryland eleven times in one year and spent $633 million. Specifically at Green Ridge State Forest, a recent study in 2008 showed that overnight recreation contributes $12.5 million per year to the economy.

The economic values of timber harvests and ecosystem services, such as the value of clean water that forests provide, have not been mentioned. Since its beginnings as a two-thousand-acre state forest in 1931, Green Ridge State Forest has made important contributions to society, economy and the environment. Elected officials from both county and state government, along with countless conservation-minded individuals and volunteer-oriented organizations, have worked tirelessly to make Green Ridge State Forest what it is today: a cultural and natural treasure, one of Maryland's crown jewels. After recently celebrating the centennial of Maryland Forestry and Parks, we also celebrate the vision and hard work of frontier and forestry pioneers who made this great forest a reality.

# PIONEERS WALK THIS LAND

*You are engaged in pioneer work.*
*—President Theodore Roosevelt*

On March 26, 1903, President Theodore Roosevelt gave the address "Forestry and Foresters" at the Washington, D.C. home of Gifford Pinchot, "the father of American Forestry." This address, later published in *Proceedings of the Society of American Foresters* 1, no. 1 (May 1905), was given to the Baked Apple Club, consisting of foresters who were part of the newly founded Society of American Foresters. Fred W. Besley, later to become Maryland's first state forester and founder of Green Ridge State Forest, was a member of the Baked Apple Club and was present on this day to hear President Roosevelt's inspiring message:

> *I believe that there is no body of men who have it in their power to-day to do a greater service to the country than those engaged in the scientific study of, and practical application of, approved methods of forestry for the preservation of the woods of the United States...*
>
> *You yourselves have got to keep this practical object before your minds; to remember that a forest which contributes nothing to the wealth, progress, or safety of the country is of no interest to the Government, and should be of little interest to the forester. Your attention must be directed to the preservation of the forests, not as an end in itself, but as a means of preserving and increasing the prosperity of the nation...*

*You must convince the people of the truth—and it is the truth—that the success of home makers depends in the long run upon the wisdom with which the nation takes care of its forests. That seems a strong statement, but it is none too strong...*

*"Forestry is the preservation of forests by wise use," to quote a phrase I used in my first message to Congress. Keep before your minds that definition. Forestry does not mean abbreviating that use; it means making the forest useful not only to the settler, the rancher, the miner, the man who lives in the neighborhood, but, indirectly, to some great river which has had its rise among the forest-bearing mountains...*

*As all of you know, the forest resources of our country are already seriously depleted. They can be renewed and maintained only by the co-operation of the forester with the practical man of business in all his types, but above all, with the lumberman...I cannot too often say to you...that you must keep your ideals high and yet seek to realize them in practical ways. The United States is exhausting its forest supplies far more rapidly than they are being produced. The situation is grave, and there is only one remedy. That remedy is the introduction of practical forestry on a large scale, and of course, that is impossible without trained men, men trained in the closet, and also by actual fieldwork under practical conditions.*

*You have created a new profession of the highest importance, of the highest usefulness to the State, and you are in honor bound to yourselves and the people to make that profession stand as high as any other profession, however intimately connected with our highest and finest development as a nation. You are engaged in pioneer work in a calling whose opportunities for public service are very great. Treat that calling seriously; remember how much it means to the country as a whole...*

*The profession you have adopted is one which touches the Republic on almost every side—political, social, industrial, commercial; to rise to its level you will need a wide acquaintance with the general life of the nation, and a viewpoint both broad and high...You have a heavy responsibility...for upon the development of your work the development of forestry in the United States and the production of the industries which depend upon it will largely rest. You have made a good beginning, and I congratulate you upon it.*

# A MARYLAND MAN WHO MATCHED THE MOUNTAINS

## THOMAS CRESAP: THE PATHFINDER

*We should protect the cultural resources at colonial Oldtown Village and designate it as a heritage area. It is an uncut diamond and compares in cultural significance to the colonial villages at Jamestown, Virginia, and Plymouth, Massachusetts.*
*—Jilla Allen Smith, president of the Irvin Allen/Michael Cresap Museum*

Maryland's great pathfinder, pioneer and patriot, Thomas Cresap, blazed the first roads through many of Maryland's present-day state forests and state parks, including Green Ridge State Forest. What Daniel Boone is to Kentucky and Davy Crockett is to Tennessee, Thomas Cresap is to Maryland.

The recent events celebrating the two-hundred-year anniversary of the Lewis and Clark expedition (1803–6) remind us of the legendary American "pathfinder." Frontiersmen like Lewis and Clark symbolize the American spirit and character. Displaying "dauntless courage," they risked it all to explore and blaze trails into the remote American wilderness. They opened a "gateway to the west," benefiting thousands of westward-bound pioneers who built and made the nation strong.

Colonel Thomas Cresap's grave site at Oldtown, Maryland. *Courtesy Tom Darden.*

The bicentennial celebrations of Lewis and Clark's expedition bring to mind two other great American frontiersmen and pathfinders: Kentucky's Daniel Boone (1734–1820) and Maryland's Thomas Cresap (1694–1787). Boone blazed a 170-mile section of the Wilderness Trail from Kingsport, Tennessee, through the Cumberland Gap of Virginia and into Kentucky (1769). Cresap blazed the Nemacolin Path/Braddock Road through the Cumberland Narrows in western Maryland, into the Ohio Valley and beyond (1753). It became "the National Road: the road that built the nation." Historian Archer Butler Hulbert wrote that the remarkable series of events that occurred along this road "mark this mountain thoroughfare as the most historic, perhaps, on this continent."

*The Pathfinder*, an 1841 novel by James Fenimore Cooper, popularized the word "pathfinder," which soon thereafter became a fixed part of the American language. The hero of the story is said to be modeled after Daniel Boone; however, Thomas Cresap might have served just as well as Cooper's inspiration. Cresap's achievements certainly rivaled Daniel Boone's accomplishments.

Consider the more than two hundred miles Cresap blazed during his twenty-six years as a pathfinder. It was a huge achievement. During the years 1733–59, Cresap cleared the first important colonial roads west from Baltimore, Maryland, to Pittsburgh, Pennsylvania.

In 1733, Cresap blazed the Old Conestoga Road, later known as the German/Monocacy Road. Originally, this trail connected the Indian village of Conestoga near present-day Lancaster, Pennsylvania, to the Indian village of Opequon near present-day Winchester, Virginia. The Conestoga Indian trail passed through today's "golden mile" on Route 40 in Frederick, Maryland, and through Fox's Gap at South Mountain State Park. Cresap cleared about fifty-three miles of the Old Conestoga Road beginning at his residence, in an area the Indians called *Conejohela*. From Cresap's home on the west bank of the Susquehanna River four miles south of Wrightsville, the road continued west through the present towns of York, Littlestown and Hanover in Pennsylvania and near Taneytown and Union Bridge in Maryland. This was the first colonial road to open west of the Susquehanna River into the Maryland and Pennsylvania wilderness.

In 1734, Cresap blazed a trail from Conejohela to Rock Run, which flows into present-day Susquehanna State Park near Lapidum, Maryland, upriver from Havre de Grace. Starting from Rock Creek, Cresap cleared the road about thirty-four miles paralleling the west bank of the Susquehanna River north to his residence at Conejohela.

Cresap began his epic work as a pathfinder on the Nemacolin Path in 1753. Hulbert stated that the Nemacolin Path was "one of the most important Indian paths in America, if indeed it was not the most important, in so far as Europeans were concerned." This trail originally extended from Will's Creek to the Monongahela River. Beginning at Cumberland, Maryland, and heading west, Cresap blazed a road over Haystack Mountain and west through Frostburg, Savage River State Forest and Grantsville in Maryland. The road continued through Uniontown to Brownsville in Pennsylvania, near the village of Delaware chief Nemacolin. With the help of scout and frontiersman Christopher Gist and friend Nemacolin, Cresap cleared the first road west of the eastern continental divide and the mountain barrier called the Allegheny front into the Ohio Valley.

General Braddock's disastrous military campaign passed over this trail in 1755. The 250$^{th}$ anniversary of this event occurred in 2005. With Braddock were George Washington and Daniel Boone. In 1803, Meriwether Lewis traveled along this road to join William Clark to begin their celebrated western expedition.

In 1758, Cresap cleared a sixty-two-mile section of road between Fort Frederick and Fort Cumberland. The original eighty-mile road crossed the Potomac River twice between the two forts. This caused transportation problems, especially during flooding. Cresap's realignment of the road, paralleling the Maryland side of the Potomac River, was a significant improvement.

Today, this road begins at Fort Frederick State Park and extends westward past the towns of Hancock and Little Orleans. It continues on Oldtown Road through Green Ridge State Forest, passing through Oldtown and Spring Gap before reaching Cumberland.

According to historian Kenneth P. Bailey, Thomas Cresap "did as much as any single person to further the westward movement." If a monument were constructed on the scale of Mount Rushmore to honor America's great pathfinders, certainly the image of Maryland's own Thomas Cresap would gaze over the great American landscape along with the likenesses of Daniel Boone, Meriwether Lewis and William Clark.

## Thomas Cresap: The Patriot

*He left his mark across the land. He's the pioneer, patriot, the pathfinder-man. What Boone is to Kentucky, Cresap is to Maryland.*
*—from the song "Ballad of Thomas Cresap: The Frontiersman"*

A little more than 250 years ago, on May 8–9, 1755, General Braddock's army camped on Thomas Cresap's fields at Oldtown, Maryland. As one of Maryland's leading colonial patriots, Colonel Thomas Cresap had an uncanny knack for being involved with many of the remarkable events and important people of his time.

One of Braddock's key objectives during the French and Indian War was to force France to abandon its claims to the Ohio Valley.

Navigating the Nemacolin/Braddock Trail that Cresap had blazed in 1753 from Fort Cumberland to the Monongahela River, Braddock marched to Fort Duquesne (Pittsburgh), the first enemy target on Braddock's list. Traveling with General Braddock's military expedition were George Washington, Daniel Boone, Christopher Gist and John Walker, all of whom probably camped with Braddock's forces at Oldtown on Cresap's fields.

On the morning of May 8, 1755, part of General Braddock's army crossed the Potomac River at the mouth of the Little Cacapon, just south of Green Ridge Forest, from Virginia (presently West Virginia) into Maryland. Braddock's forces then followed the Potomac River upstream about nine miles to Cresap's Fort at Oldtown, arriving there early in the evening. Cresap's Fort was Braddock's last stop on a several-month journey from Virginia to Fort Cumberland. For two nights, an estimated 1,400 men from Braddock's army camped at Oldtown. Of these men, about 450 were colonial soldiers led by Lieutenant Colonel George Washington.

General Braddock spent the night in Thomas Cresap's house. Cresap, true to his nature, probably offered Braddock advice on tactics and strategies to fight the French and Indians. However, Braddock most likely ignored Cresap's counsel, just as he had disregarded Benjamin Franklin's at Frederick earlier on this trip.

At Oldtown, the paths of Thomas Cresap and Daniel Boone, a wagoneer for General Braddock, crossed. During Braddock's campaign, Boone heard stories about Kentucky and its natural wonders. Perhaps it was on Cresap's fields that Boone decided he would one day explore Kentucky. Boone would have profited with plans to explore Kentucky by talking to Christopher Gist, Braddock's chief trail guide. Cresap handpicked Gist to survey and map the upper Ohio Valley for the Ohio Company, which promoted English settlements in this region. In 1750, during this assignment, Gist penetrated the northeastern Kentucky wilderness, about eighteen years before Boone led pioneers through the Cumberland Gap.

Boone would have also benefited from talking with Dr. Thomas Walker, commissary general to Braddock's Virginia forces. In 1750, the same year as Gist's Kentucky expedition, Walker explored southeastern Kentucky and was the first to record the existence of the

Cumberland Gap. Cresap certainly would have spoken with Walker since the Maryland frontiersman also served as a commissary man. Cresap controlled storehouses along Braddock's route and provided food and supplies to his army.

Dr. Thomas Walker's neighbor in Virginia was Peter Jefferson, Thomas Jefferson's father. When Peter Jefferson died, Thomas was fourteen years old. For the next seven years, Walker became Thomas Jefferson's guardian, overseeing his growth and education to adulthood. During Cresap's lifetime, Thomas Jefferson defined the word "patriot": a person "who loves his country on its own account, and not merely for [personal] interest or power...[and] can never refuse to come forward when he finds that she [the country] is engaged in dangers which he has the means of warding off."

Cresap's life was a parade of remarkable experiences, which his role as a patriot promoted, often at great personal and economic risk. Cresap participated in four colonial wars: the Conojacular War (1730–38), the French and Indian War (1754–63), Lord Dunmore's War (1774) and the American Revolutionary War (1775–83).

The bravery Cresap demonstrated in these colonial conflicts made him famous in his own lifetime. By his senior years, he was a living legend. People went out of their way to Oldtown to meet and talk to the "celebrated Colonel Cresap." They wanted to hear firsthand the colorful stories directly from the "bantam sized Indian fighter," as author Allan W. Eckert described Thomas Cresap, who "had so much courage that it seemed to seep out of his very pores."

They may have heard Cresap tell robust tales about Braddock, how he was defeated at the Battle of the Wilderness on that summer day in July 1755, or heard Cresap's reflections of when, on August 2, 1755, the remnants of Braddock's army camped one last time in Cresap's fields at Oldtown before returning to Virginia.

## Thomas Cresap: The Pioneer

*A pioneer is someone who goes into an unknown land to carve out a new life for themselves and their families.*
*—Cathy Clark, Kentucky park manager*

# A Maryland Man Who Matched the Mountains

What is a pioneer? Dean Henson, naturalist and cultural history interpreter with the Kentucky Department of Parks, said this about colonial pioneers: "The pioneers of our frontier era were people who matched the mountains; a stout, self-reliant, utilitarian people. Taking great pride in being stalwart individuals, they went forward fully aware of the incredible risks and dangers that lie ahead. Yet, they didn't go in fear; they went in spite of it."

It must have been difficult for those early pioneers to scratch out a living. In many places of the county, where the soils are shallow and rocky, farming ground is poor. Ed George, engineer for the Maryland Department of General Services, told me recently that in areas like this, the land was so poor for growing things that "you couldn't even raise heck with a gallon of whiskey."

*The Pioneers* (1823), authored by James Fenimore Cooper, introduces the fictional character Natty Bumppo. Critics praise it as the first great American novel. Following this, Cooper wrote a series of books collectively known as "the Leather Stocking Tales." They highlight the adventures of Bumppo at various stages of his life on the American frontier. According to historians, Cooper may have modeled the main character of his idolized frontier hero, Bumppo, after Daniel Boone. However, Thomas Cresap could have served as Cooper's inspiration just as well. Like Bumppo, both Boone and Cresap were "pathfinders, pioneers, and patriots." In addition, all three strongly believed in individual rights. Like Bumppo, Boone and Cresap displayed extraordinary wilderness survival skills, requiring an intimate knowledge of the natural world, especially the flora and wildlife of the Appalachian forests; and, like Bumppo, they mastered the long rifle and were elite sharpshooters.

However, in one particular way, Cresap seems to model Bumppo more so than Boone. Bumppo shared his frontier adventures with a Native American friend and companion named Chingachgook of the Delaware Nation. Bumppo and Chingachgook are linked together forever in literature. Likewise, Thomas Cresap and Nemacolin are linked together forever in history. Nemacolin, Cresap's lifelong friend, was also a Native American of the Delaware Nation. Tradition and documents tell us that Nemacolin, with frontiersman Christopher Gist, helped Cresap blaze the Nemacolin Path. Today, this well-traveled corridor is famous as the National Road, "the road that built the nation."

Jim Ettman, a Kentucky Park naturalist, stated, "A pioneer is anyone who steps out of the box, takes a risk, and leaves their comfort zone." Cresap certainly met this definition. For most of his life, Cresap and his family lived outside the safety zone of civilization in the most extreme reaches of the Maryland frontier. Documents reveal that between 1730 and 1755, Cresap was Maryland's foremost pioneer, leading the way for thousands of pioneers who followed; they claimed and secured the land for Lord Baltimore and Maryland.

In 1710, around the age of fifteen, Thomas Cresap immigrated to America from Skipton, England, where he was born. He initially settled in the area near Havre de Grace, Maryland. By the late 1720s, Cresap was operating Cresap's Ferry several miles upriver on the Susquehanna River, between present-day Fort Deposit and Lanham, Maryland.

The years between 1730 and 1742 were extremely active and eventful. During this twelve-year period, the Cresaps moved three times: in 1730, to Conejohela (near present-day York, Pennsylvania); by 1739, to present-day Williamsport, Maryland; and by 1742, to Oldtown, Maryland. At Conejohela and Oldtown, the Cresaps were the first Maryland family to settle in the region.

Nina Higgins, a direct descendant of Thomas Cresap, recently shared this definition of a pioneer: "The pioneer lights a torch, foresees great visions for the future, acts on those visions, and then passes that lighted torch down through the generations."

# THE CARROLL FAMILY: SHAPING HISTORY IN AMERICA AND AT GREEN RIDGE

*Without morals a republic cannot subsist any length of time; they
therefore who are decrying the Christian religion…are undermining
the solid foundation of morals, the best security for the duration of free
governments.*
—*Charles Carroll, signer of the Declaration of Independence and early
landowner of the Green Ridge Forest area*

The early pioneers of Green Ridge State Forest made important
contributions to the growth of our nation. Colonel Thomas
Cresap and George Washington are two prominent figures that
often come to mind that have connections to GRSF. The Carroll
family also had solid ties to Allegany County, particularly GRSF,
and has been another great influence in shaping the history of
our country.

In the early 1800s, Charles Carroll of Carrollton—the last
surviving signer of the Declaration of Independence, Maryland
member of the Continental Congress and the only Roman Catholic
to sign the Declaration of Independence—owned most of present-
day Green Ridge State Forest. He was one of the wealthiest men
of his time. He also served as U.S. senator for Maryland (1789–92).
The Wellesley Hollow Trail and Stafford Road are place names at
GRSF named after Carroll's grandchildren.

Daniel Carroll of Rock Creek was a framer of the U.S.
Constitution and cousin of Charles Carroll. In the early 1800s,

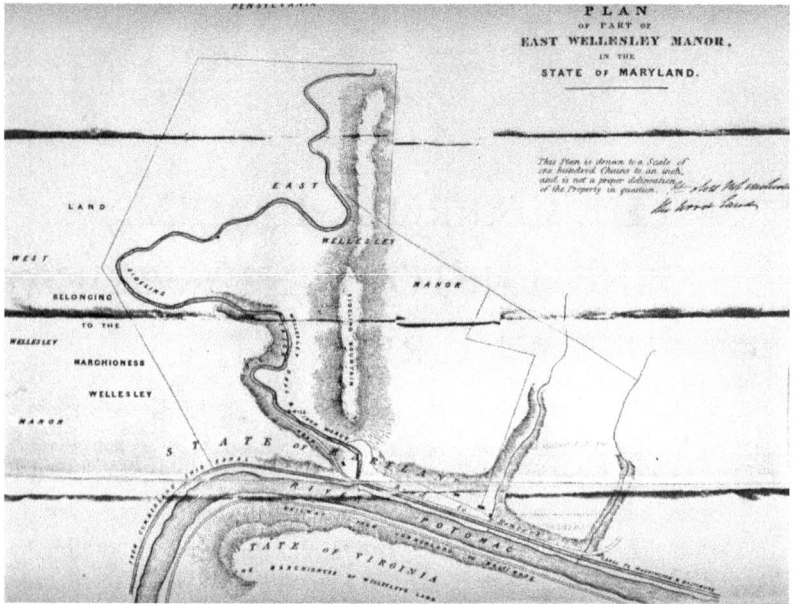

An 1840 map of the land owned by Charles Carroll's granddaughter, Lady Wellesley, that extends east from Green Ridge forest into adjacent Washington County. *Courtesy Maryland Historical Society.*

William Carroll, Daniel Carroll's grandson, with Richard Caton, son-in-law of Charles Carroll, oversaw the business operations at Green Ridge with a special interest in iron ore and timber. In 1836, Charles Carroll's estate financed the building of the Carroll's Steam Saw Mill, which included a gristmill, blacksmith shop, stables and quarters for workers. The Carroll Chimney, a part of the steam-powered sawmill, is the only surviving structure from this period.

On November 6, 1789, Pope Pius VI appointed as the first bishop in the United States Father John Carroll, cousin of Charles Carroll. John Carroll once held the deed to the land where St. Patrick's Catholic Church currently sits at Little Orleans on the southeast side of GRSF. The church represents the earliest continuous rite of Catholic worship in western Maryland.

Charles Carroll helped guarantee religious freedom in Maryland by assisting in the writing of the Maryland State Constitution. He also assisted the bishop in advocating for the United States Bill of Rights.

Celebrate with joy for our freedoms and liberty. Take a ride on Carroll Road, one of the oldest roads in eastern Allegany County, visit the ruins of Carroll Chimney and enjoy the sublime, spectacular scenery from Point Lookout.

# Building Ships and Utopia in the Maryland Mountains

*They are living in apple paradise and do not know it.*
*—Professor H.E. Van DeMan, promoter of Green Ridge Valley Orchards*

Frederick Mertens (1824–1886), who came from a long family line of shipbuilders, immigrated to America from Hamburg, Germany. After spending a bit of time in South America, New York and Hollidaysburg, Pennsylvania, he visited Cumberland, Maryland, in 1851, eventually permanently settling there in May 1852. On April 14, 1856, having resided in the United States for five years, he became a citizen of the United States.

You may be asking yourself, *Why would a shipbuilder move to the mountains of western Maryland?* The answer: to build canalboats for the Chesapeake and Ohio Canal. The construction of the 184.5-mile canal, an engineering marvel, had just been completed from Georgetown to Cumberland in 1850.

A person who could build boats economically and efficiently to operate on the C&O Canal could do very well in business. In fact, Mertens amassed a fortune, and by the 1870s, he was considered one of the wealthiest men in Cumberland, having expanded his business from shipbuilding to the coal and lumber business, as well as managing Queen City Glass Works. He also began acquiring large tracts of forested land to support his lumber business. One large tract of land over thirty thousand acres in size was acquired

Mertens Shipyard and Wharf, Cumberland, Maryland, circa 1880–1910, from Mertens photo album. *Courtesy Tom Hynson.*

Dam at C&O Canal in Cumberland, Maryland, circa 1880–1910, from Mertens photo album. *Courtesy Tom Hynson.*

Point Lookout at Green Ridge, circa 1880–1910, from Mertens photo album. *Courtesy Tom Hynson.*

Point Lookout at Green Ridge, circa 1880–1910, from Mertens photo album. Note the water in the C&O Canal. *Courtesy Tom Hynson.*

from the heirs of Charles Carroll of Carrollton in eastern Allegany County. This tract of land later came to be Green Ridge State Forest.

Frederick Mertens and his wife, Mary (1835–1904), had five sons and a daughter: John, Henry, Edward, Frederick, William and Emma. Two of their sons, Frederick and John, were the primary members of the family involved in the business management of their landholdings in eastern Allegany County.

John Mertens married Elsie Yeargan, a direct descendant of William Yeargan, who came "from Prussia Germany and settled early in his youth in Cumberland Maryland and learned the boat builders trade and helped to construct the first boat used on the Chesapeake and Ohio Canal." In the late 1860s, Yeargan purchased the Samuel Sonnaborn farm located at present-day Rocky Gap State Park. The farmhouse was located near the site of present-day North Beach at Rocky Gap. Yeargan's obituary, printed in the *Cumberland Evening Times* of June 1, 1908, stated that Yeargan improved his property to a point that it was considered one of the best farms in Allegany County.

*Above*: City of Cumberland, Maryland, circa 1880–1910, from Mertens photo album. The Queen City Hotel is in the center left of the picture. *Courtesy Tom Hynson.*

*Right*: Frederick Mertens (1824–1886) established a canalboat-building business in Cumberland, Maryland. *Courtesy Tom Hynson.*

Mary Julia Mertens
(1835–1904), wife of
Frederick Mertens.
*Courtesy Tom Hynson.*

Frederick Mertens Sr. purchased General McCaig's mansion on April 3, 1876. The mansion was located on Baltimore Street on the site of the present-day YMCA building in Cumberland. Just as Richard Caton's mansion at Thunderhill in Catonsville, Maryland, served as a long-distance office to run Charles Carroll's iron ore and timber enterprises at Green Ridge, likewise, Mertens's mansion in Cumberland served as a long-distance office to run his vast business empire, including the timber and fruit orchard business at Green Ridge. On April 20, 1880, Fred Mertens connected a telephone line from his office to the C&O Canal that ran from Cumberland to Georgetown. At this time it was one of the earliest and longest telephone lines in the country. This early use of technology allowed Mertens to track the progress of all his boats operating along the 184.5-mile canal.

Each year in the spring, Mertens launched newly constructed boats from his Cumberland boatyard that employed seventeen men or more. In February 1874, Mertens launched twelve new boats on the canal. By the 1870s, people called the wealthy capitalist

The Mertens Mansion at the intersection of Baltimore Avenue and Baltimore Street was razed about 1925 to construct the YMCA building. *Courtesy Tom Hynson.*

The Mertens office next to the Mertens Mansion, Baltimore Street, Cumberland, Maryland, circa 1880–1910, where Green Ridge business occurred. The child on the steps is Fred Mertens III, grandfather of Tom Hynson, contributor of the Mertens photo album pictures. *Courtesy Tom Hynson.*

"Captain Mertens" because of all the boats he owned and operated. Boats owned by Mertens came with names such as the *Elbe, J.W. Turner, B.F. Price, W.P. Wools, F.A. Mertens, Spier, Clara, G.H. Bradt* and the *Belle Besley* (the final name a foreshadowing of what was to come, as Fred Besley, Maryland's first state forester, beginning in the 1930s, began purchasing for Green Ridge State Forest many of Mertens's orchard lots).

In April 1894, Mertens was not too busy with business to provide assistance to Coxey's army when it arrived in Cumberland. Coxey's army was en route to Washington, D.C., to plead with lawmakers to provide jobs for many who were out of work because of the great economic depression that began in 1893, the worst economic crisis to hit the United States at that time in history. The men of this army wanted the government to fund road-building jobs to put people back to work. Historians record this event as the first significant protest march in America's history. Coxey's army began its march at Massillon, Ohio. After spending two days in Cumberland, on April 17, 1894, the men left by canalboats, many of them belonging to Mertens, and floated on the canal past Green Ridge on their way to Washington, D.C.

The Mertenses' dining room boat on the C&O Canal near Green Ridge Forest. *Courtesy Tom Hynson.*

View of the Mertenses' boatyard in Cumberland, Maryland, circa 1880–1910. *Courtesy Tom Hynson.*

Frank L. Baum, not yet a famous writer, observed Coxey's army from a distance. The march must have made an impression on him. Political interpretations of his book *The Wizard of Oz*, which he wrote in 1900, imply that Coxey's army influenced Baum when he wrote this classic story. The scarecrow represents a farmer, the tin woodman an the industrial worker and the cowardly lion represents a political leader. The city of Oz represents Washington, D.C. Dorothy's shoes (silver) walked on the yellow brick road (gold); these symbolize the United States' monetary policy at that time and the ongoing controversy of the use of free silver over the gold standard. So there you have it—the Mertenses and GRSF played a small part in the classic story *The Wizard of Oz*.

In the 1880s and 1890s, the Mertenses cleared and burned off much of the original forest for timber at Green Ridge. Not long after Frederick Mertens's death in 1886, his sons began developing grandiose ideas for a new business venture. Their plans involved

Baltimore Street, Cumberland, Maryland, circa 1880–1910, from Mertens photo album. *Courtesy Tom Hynson.*

City of Cumberland, Maryland, circa 1880–1910, from Mertens photo album. *Courtesy Tom Hynson.*

Note the youth of two of the Mertens lumbermen, circa 1910, Mertens photo album. *Courtesy Tom Hynson.*

John Mertens, with his brother Fred Mertens, managed much of the family business activity at Green Ridge, circa 1900. *Courtesy Tom Hynson.*

establishing the largest orchard, not in the United States, not in the world, but the largest apple orchard in the "universe," as stated in a July 30, 1913 Cumberland newspaper. The model that inspired them was the successful apple orchards operations located at Hood River Valley in Oregon.

Beginning about 1895, the Mertenses began a fifteen-year study regarding the establishment of a commercial apple orchard industry in the Green Ridge area. Experts made it sound like Green Ridge was the Garden of Eden for growing apple trees: "In open competition with the best apples grown in America, both in exhibition and in markets, the apples of western Maryland have proven their superiority."

The Mertenses appeared to be familiar with George Sudworth's Allegany County 1900 forest inventory, for they cite the same number of tree species and point out the incredible tree species diversity found in the Green Ridge region.

*Situated as it is between the North and the South, on the borderland just south of the Mason and Dixon's Line, Green Ridge Valley*

Mertens Green Ridge Logging Railroad, circa 1880–1910, with John Mertens at left. *Courtesy Tom Hynson.*

Mertens lumber team at Green Ridge Station at the present-day intersection of Kasekamp Road and Mertens Avenue, circa 1910. *Courtesy Tom Hynson.*

Mertens lumber camp, circa 1910, near Green Ridge Station. *Courtesy Tom Hynson.*

Cumberland, circa 1880–1910, Mertens photo album. *Courtesy Tom Hynson.*

*Partakes in a peculiar way of the features of both the boreal [northern] and the austral [southern] of the flora of North America. Consequently, there is a richness of species which neither the higher lands to the west, more boreal, nor the lower lands to the east, more austral, exhibit in richness of growth or in number of species. The sides of the hills in the valley are covered with a second growth and with some few remnants of the once great wealth of lumber trees which clothes their sides. Even in this condition there have been observed within the general area no fewer than seventy-two species of conifers and hardwood trees in marked intermingling of north and middle south. The exuberance of tree growth is of greatest significance in connection with the tree growing qualities of the area. When it is further noted that the apple and kindred fruit families [hawthorne] are included among the sixty-four species of hardwoods growing luxuriantly and abundantly over the entire area, the value of this added testimony is of profound interest. The fact that they occur over the entire area and flourish so luxuriantly amid surroundings in*

Green Ridge logging railroad run by the Mertenses, circa 1880–1910. *Courtesy Tom Hynson.*

John Mertens trying to make the sale of a ten-acre orchard lot at Green Ridge, circa 1910. *Courtesy Tom Hynson.*

Construction of Mertens's garage for his car—one of the first seen in the area by Green Ridge residents, circa 1910. *Courtesy Tom Hynson.*

*no way contributing to their well-being, is confirmatory evidence*
*of Green Ridge Valley being naturally acceptable to orcharding.*

Mertens certainly could state that Green Ridge Valley had excellent access to major markets on the East Coast. The big cities were within easy reach of Green Ridge because of the Baltimore Pike, a first-class road, as well as "excellent transportation facilities over the Baltimore and Ohio Railroad, Western Maryland Railroad, and the Chesapeake and Ohio Canal." In another promotional pamphlet, Professor H.E. Van DeMan wrote in 1910 that there was no place on earth out of reach of the Green Ridge Valley Orchard: "The markets of the world lie at the foot of their mountain peaks. Not only are the cities of the eastern and central states within less than a day's run of an ordinary freight train but the great ports are likewise there, whose thousands of ships reach the very ends of the earth. The trade in American apples has only begun, both at home and abroad."

The Mertenses were planning to build a small city at Green Ridge: "We plan to build up a town of commercial importance. The largest possible development of the Green Ridge Valley apple industry and

Unloading crates at Green Ridge Station at the present location of Mertens Avenue and Kasekamp Road, circa 1910. *Courtesy Tom Hynson.*

the building of the town of Green Ridge are both of great interest to us, and we desire to co-operate fully in all plans looking to that end." To accomplish this, the Mertenses divided the Green Ridge Valley orchard property into more than three thousand lots, each about ten acres in size. The Mertens sons planned to establish town sites within the orchard. Old maps made by the Mertenses show plans for the "villa sites." One town site was located on Town Hill near No Name Lookout; another was on Polish Mountain north of present-day Interstate 68; and one was at the foothills of Town Hill near the intersection of Mertens Avenue and Oldtown Road.

The Mertenses stated, "Green Ridge Valley is of such magnitude that we have subdivided a portion of it for the purpose of developing thereon a large fruit-growing community and to build up at the same time a town of commercial importance." The Mertenses designed the town sites so "that all owners of apple orchards in Green Ridge Valley could participate as property owners in the town of Green Ridge, with an opportunity to maintain a home in the town, if that is preferred to residence upon the orchard property."

For about a century, since the Charles Carroll and Mertens families owned the land, Green Ridge had been kept as one large

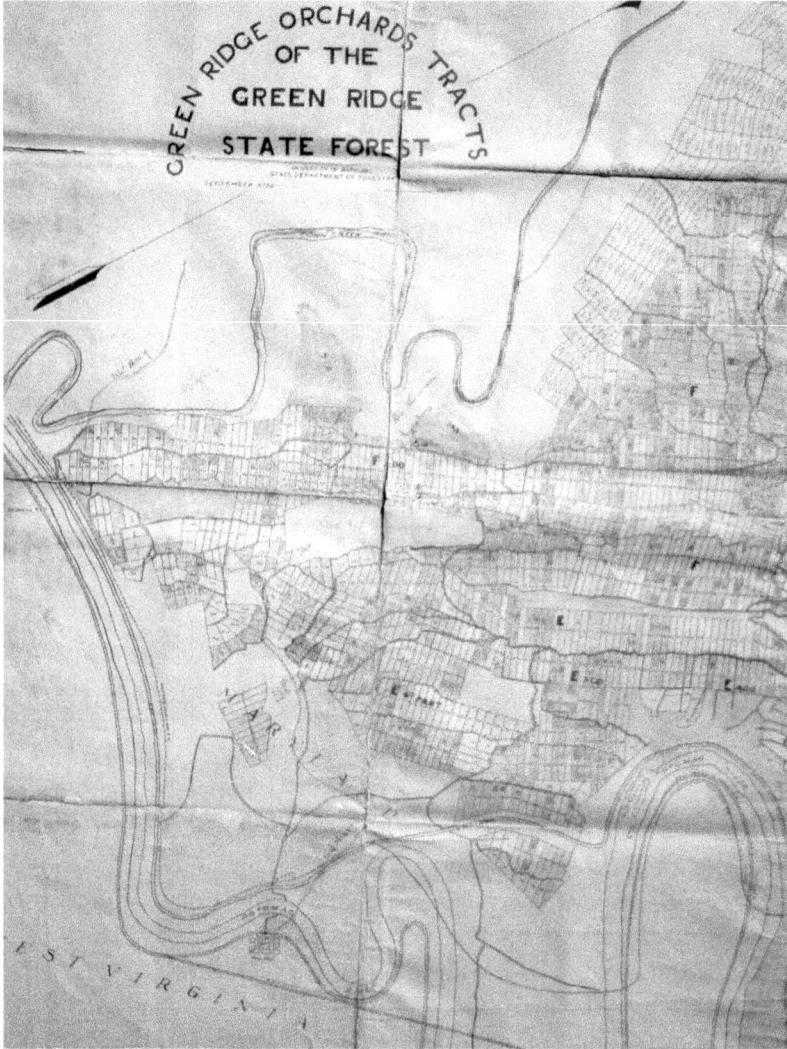

A portion of a Green Ridge Valley Orchard map showing property subdivided into many ten-acre orchard lots. *Courtesy Maryland Department of Natural Resources.*

tract of land; however, in 1910, that was about to change now that the Mertenses planned to subdivide the land into ten-acre lots and sell them to individuals across the country. The Mertenses wrote: "We intend to dispose of our property...based upon our conviction that a large community will attract to Green Ridge a substantial population, which will be highly advantageous to all business interests, and will inevitably cause a much greater enhancement of all land values, both orchards and town property."

If they established the town sites, the Mertenses believed that people would come, and come they did. The Mertenses promoted it with statements like this:

> *Green Ridge Valley is noted for its magnificent scenery and is a delightful place for a summer residence. It has a mild and healthful climate, and it is an ideal spot for the home at all seasons of the year. It is in the midst of the richest and most populous section of the United States, and is within a few hours distance of all principal points in the eastern and central states. Its exceptional transportation facilities make the town and the valley easy of access from all points, and gives the community great advantages in a commercial way.*

The Mertenses informed prospective orchard lot buyers that "huge profits can't be doubted" from their investment. They advised prospective buyers that if they planted fifty apple trees on each acre,

Queen City Hotel, Cumberland, Maryland, circa 1880–1910, Mertens photo album. *Courtesy Tom Hynson.*

they could realize an annual net profit of $375 per acre from the sale of apples. However, they said that was a conservative estimate, as a nearby commercial orchardist earned as much $1,500 per acre in a single season from the sale of his apples.

According to the Mertens philosophy, money does grow on trees…in the form of apples! Once the trees were planted, it would not be long before they would begin bearing fruit and money. In their promotional material, the Mertenses noted a comment made in 1906 by W. McCullough Brown, one of the founders of the Maryland Forest Service and a well-known apple grower in western Maryland: "At the end of four years from planting, many of the trees will begin to bear, and in six years a considerable crop may be expected. A well-grown tree will yield twenty bushels of apples." At the going rate of three dollars a bushel, each tree would produce for the owner sixty dollars of revenue a year.

The Mertenses wrote about the value of the land itself. New orchard land without trees in other parts of the country would sell from $20 to $100 per acre. In Pennsylvania, it was difficult to purchase land with a fully matured orchard for $800 per acre, although if the truth were known, according to the Mertenses, the average price of good orchard land was $1,000 or more per acre.

The Mertenses offered investors a ten-acre tract of land for anywhere from $1,500 to $2,500, depending on its location. This was considered a bargain when compared with Pennsylvania orchard land. The Mertenses especially made out, since they had bought the land originally for about $1 per acre. The ten-acre orchard tracts were generally laid out in rectangles 300 feet wide and 1,800 feet long. With each orchard tract came a villa site, about 10,000 square feet, where one could build a permanent home or a summer cottage. The Mertens brothers would even sell the new owners lumber from their own lumberyard and build the home if the customer desired their services.

John Mash, in his book *The Land of the Living*, wrote about the utopian society the Mertenses planned to create for Green Ridge Valley members. Telephones would connect all members to the outside world and throughout the Green Ridge Valley to all the cottages, commissaries and camps. The telephone would make it

Mertens lumber camp at Green Ridge, circa 1910, Mertens photo album. *Courtesy Tom Hynson.*

Western Maryland Railroad Bridge, Cumberland, Maryland, circa 1910. *Courtesy Tom Hynson.*

possible to have groceries and supplies delivered to cottages three days a week. The association would also establish schools and a law enforcement agency.

As far as the cultivation of the apple orchards, the association would take care of everything, as well as advertise and transport the fruit to market. In the meantime, while the money rolled in, the orchard owners could relax at the White Sulphur Springs Resort once it was constructed near the hollow of present-day Gordon Road. The Carroll family had planned to develop a similar resort in the same location eighty years earlier, modeled after the White Sulphur Springs resort in present-day West Virginia. Mash wrote of this utopian society:

> *With the Association taking on the burdens of cultivation and business, the tract owner could devote himself to leisurely recreation...The owner could soak in the refreshing waters of the springs or take in mountain climbing, fishing, boating, hunting and golfing to wile away the hours. The community was*

Viaduct of B&O Railroad, Cumberland, Maryland, circa 1880–1910, Mertens photo album. *Courtesy Tom Hynson.*

Mertens Green Ridge labor crew eating lunch, circa 1910, Mertens photo album. *Courtesy Tom Hynson.*

*planned for the elite. The fellow next door would be like yourself, accustomed to the finer things in life. All owners would be the professional class—doctors, lawyers…no riffraff allowed here! The Hood River Valley community boasted about its University Club the Owners formed. Surely, the Green Ridge Valley Orchard Association would equal, or exceed anything the crude westerners could accomplish.*

In 1913, the Mertenses stepped up their promotion of the Green Ridge Valley operation. Newspaper reporters spoke of Green Ridge's unrivaled beauty and history. They wrote of how the land had come down through the family of Charles Carroll to Frederick Mertens. They noted how more than eight hundred people, primarily new immigrants to America, worked in thirty separate camps every day at Green Ridge.

A reporter wrote in a July 30, 1913 Cumberland newspaper article that the Mertenses entertained at the Green Ridge Station Hotel officials from the Third National Bank. After the officials were

The Roby family, like many farmers in the area, sold their farms to the Mertens family. *Courtesy Tom Hynson.*

The Mertens lumber camp at Green Ridge Station at the present-day intersection of Mertens Avenue and Kasekamp Road, circa 1910. *Courtesy Tom Hynson.*

*Above*: The Mertens lumber team near Green Ridge Station, circa 1910. *Courtesy Tom Hynson.*

*Right*: Chauffeur for Mertens staff at the Green Ridge Orchard, circa 1910. *Courtesy Tom Hynson.*

*Above*: Mertensville is now the site of the White Sulphur Pond area at Green Ridge, circa 1900. *Courtesy Tom Hynson.*

*Left*: Fred Mertens (right) trying to make a sale of an orchard lot at Green Ridge to prospective buyers, circa 1910. *Courtesy Tom Hynson.*

wined and dined, they were likely taken to Point Lookout, where the Mertenses gave the bankers a convincing sales pitch as they looked over the spectacular landscape. Potential orchard lot buyers were also entertained at the Queen City Hotel in Cumberland. The article mentioned that at the time the Mertenses had for sale 3,000 ten-acre tracts, of which 2,600 were sold.

After 1913, the glowing reports of the Green Ridge Valley orchard operation became less frequent. The reports really went south for the Mertens orchard enterprise early in 1917, when they were forced to declare bankruptcy. Several newspaper accounts in February 1917 mentioned that the Mertens firm had once controlled assets valued in the millions of dollars, including "timberlands, mines, steamship interest"; in reality, they were experiencing financial difficulty. A newspaper article mentioned that the Mertenses had bought thirty thousand acres of land in Allegany and Washington Counties and had divided it into lots and sold them to more than 2,500 individuals living in other cities across the country; however, they were now in big financial and legal trouble. In early 1917, many of these orchard lot investors

Mertens specialty mill, circa 1910. *Courtesy Tom Hynson.*

Tom Hynson, direct descendant of the Mertens family, at the Carroll Chimney at Green Ridge, a steam-powered sawmill built in 1836. Photo taken in 2002. *Photograph by author.*

claimed that they had paid the Mertenses their money more than a year before and had not yet received deeds for their property. Others found that when they investigated their orchard, they learned that another person claimed the same property. The Mertenses proclaimed that they were just "farmers, and as tillers of soil" they should not be forced into involuntary bankruptcy proceedings. The courts ruled differently, and it wasn't long after that the Green Ridge Valley Orchard enterprise was declared bankrupt.

Fred and John Mertens left Cumberland and built up a tourism business operating steamboats on the Potomac River that took passengers from Marshall Hall to Mount Vernon and back, just downstream from Washington, D.C. In a way, the Mertenses, after abandoning their fruit-growing enterprise at Green Ridge, returned to their ancestral occupation involving ships and boats, like their shipbuilding ancestors from Hamburg, Germany. In their minds, they might have thought they were completely divorced from Green Ridge, but this was not really true, for there was still an uninterrupted

Steamer "Charles Macalester" for Mt. Vernon, Capacity 1600

After the Mertenses' failed business venture at Green Ridge, they hosted tourists on the lower Potomac River, transporting passengers from Washington, D.C., to Marshall Hall to Mount Vernon and back on the steamboat, the *Charles Macalester*. *Courtesy Tom Hynson.*

hydrological thread that kept them connected to the orchard land in Allegany County: the source of some of the water on which their steamships floated came many miles upstream from the watershed of Green Ridge Forest.

# LAND OF FORESTS AND TREES

*The forests of Allegany County are one of her chief assets.*
*—Fred W. Besley, 1912*

An estimated 95 percent of Maryland's landscape was covered in forest when Lord Baltimore's first settlers arrived at St. Mary's on the ships *Ark* and *Dove* in 1634. Today, forests still cover about 41 percent of Maryland. This is remarkable when one considers that Maryland is the fifth most densely populated state in the United States, with an estimated 5.8 million people in 2010.

Of the twenty-three counties in Maryland, Allegany County has the highest percentage of forest cover, with 78 percent, while Garrett County, its western neighbor, follows closely in second place with 72 percent forest cover.

The 43,560-acre Green Ridge State Forest makes up about one-fifth of the Allegany County land base, the largest contiguous block of forestland in Maryland within the Chesapeake Bay watershed. GRSF is a part of the central hardwood forest region of the United States that extends along the Appalachians north and south from New York to Georgia and east to west from western Maryland to Missouri.

America's central hardwood forests, of which GRSF is a part, contain the largest mixture of deciduous trees found anywhere in the world. Foresters often classify the forests at GRSF as an oak-hickory forest. This understates the incredibly rich diversity of tree species found in this area. Rutherford Platt, in his book *The Great American Forests*, called the Appalachian region of hardwoods "the gayest, most colorful, most livable and bountiful forest in the world."

One really begins to appreciate the rich diversity of Green Ridge when comparing it to the diversity of European forests. For example, the United Kingdom—consisting of England, Scotland, Wales and Northern Ireland—has about half the number of native tree species growing over its territory of 94,526 square miles as Green Ridge State Forest has in less than 200 square miles. Forest inventories that occurred over the past century in Allegany County and GRSF detail the rich plant diversity of the area.

# SCIENTIFIC FORESTRY COMES TO ALLEGANY COUNTY

*Trees, the largest member of the plant kingdom teams up with men and women, the most advanced member of the animal kingdom...that makes for an unbeatable combination.*
—*Adna "Pete" Bond, Maryland state forester (1968–77)*

G arrett County is rightfully called "the cradle of forestry in Maryland." Here, in 1906, the Maryland State Board of Forestry was established to manage the Garrett brothers' donation of land to the state that became Garrett State Forest, Maryland's first state forest. This event is well documented in historical records of the Maryland Forest Service; however, a lesser-known yet equally important conservation-related event took place six years earlier in Allegany County. This is where the first scientific forest inventory took place. Perhaps, therefore, Allegany County can be called "the birthplace of scientific forestry in Maryland."

Several important pioneering works of forestry occurred in Allegany County over the past century that involved forest inventories. Examining information in these inventories provides a snapshot of the condition of the forest the year the field data were collected. It also provides an appreciation of how resilient the forest really is. It is hard to imagine that the generally healthy forest landscape we know today was in a devastated condition just a little more than one hundred years ago. It is also impressive to realize that Allegany County and Green Ridge State Forest were on the leading edge of establishing new scientific forestry techniques and technological breakthroughs with the 1900 Allegany County Forest Survey.

William Bullock Clark (1860–1917), the department head of the Maryland Geological Survey, supervised the first forest survey in Allegany County in 1900. His office was located at Johns Hopkins University in Baltimore, Maryland. Clark was one of Maryland's best-known and most respected scientists of his time. In 1906, Clark supervised Maryland's first state forester, Fred W. Besley, when the Maryland State Board of Forestry was established. In 1908 and 1909, Clark oversaw and guided Besley when he conducted another forest inventory in Allegany County, which included the area of the present-day Green Ridge State Forest.

George Bishop Sudworth (1864–1927), a forester for the U.S. Forest Service, conducted the fieldwork for the 1900 Allegany County Forest Survey. He worked in Maryland under the guidance of Clark. Sudworth, a forestry conservation pioneer in his own right, wrote many publications about trees and forestry. *A Check List of Forest Trees of the United States* is perhaps Sudworth's best-known publication. During his career, he discovered and named many types of American trees across the country. By the end of his career, Sudworth was the chief dendrologist for the U.S. Forest Service.

In 1900, Sudworth wrote that the main purpose of the forest inventory in Allegany County was to develop the economic resources of Maryland. Allegany County was chosen as the place for the "beginning of this work," thus making Allegany County the place where the first scientific forestry inventory occurred in Maryland.

In 1900, Sudworth found that the forests of Allegany County, because of overharvesting, uncontrolled grazing of livestock and damaging forest fires, were in very poor condition, needing a lot of work to restore them as a viable economic resource. For the inventory, Sudworth focused his attention on the condition, composition, character and uses of the forest.

Sudworth noted that he inventoried the forests in the county by traveling on foot, rail and sometimes by a team of wagons. On his travels, Sudworth identified four conservation practices needed if the forests were to recover: protect forestlands from fire, exclude

Narrows in Cumberland, Maryland, circa 1880–1910. *Courtesy Tom Hynson.*

grazing from forestlands, regulate "indiscriminate" timber cutting and regulate timber cutting on steep slopes. Sudworth claimed that these "evil effects" wreaked havoc on forest regeneration, damaging or destroying tree seedlings. The forests were unable to reestablish themselves due to the above-mentioned damage and abuse.

In 1900, Sudworth observed that the constant pressure on the forests of Allegany County since 1840 had reduced them to a state "of the lowest productiveness, which [had] in turn left an impression among many people that this resource [was] irretrievably gone"; however, Sudworth believed that all was not lost. He was optimistic that "the rapid natural reproduction" he had observed was "most encouraging for a recuperation of these depleted forests" and that they would recover if the forests were "placed under conservative management."

Sudworth observed wasteful timber harvesting practices where loggers left up to 30 percent of the trees on site. This undesirable practice—called "high grading" today—cut the best trees and left the poor quality and unwanted scrubby trees behind, without regard for the future regeneration of the forest. Often, wildfires occurred afterward. The additional woody debris left on the site caused the wildfires to burn much hotter than they normally would, not only destroying the trees but also damaging the soil and greatly diminishing its productivity. Sudworth recommended retaining three to four seed trees of each useful timber species distributed evenly over every acre. This timber harvest method is similar to what foresters today call a selective, seed-tree harvest.

Sudworth apparently talked to some of the old-timers in the area about the original forests of Allegany County. In presettlement times, Sudworth wrote that "with the exception of a few treeless swampy meadows of small size, the entire county was once a continuous forest...the heaviest timber existed in the coves, or on the low hills, and on the lower slopes and benches of the mountains, where the soil is deepest and most porous. The rocky upper slopes and summits appear to have borne a forest of small stunted trees." Sudworth indicated that the original forest in the Green Ridge area in eastern Allegany County once produced a much larger proportion of coniferous species, such as white pine and short-leaf pine, than found today. Historic place names like Piney Grove and Piney Run indicate that large white pine and short-leaf pine once grew in the region during those pioneer days.

Sudworth reported that the timber in the original forest was large and of excellent quality, yielding eight thousand to ten thousand board feet per acre over large areas; but by 1900, the forest was only able to produce five hundred to two thousand board feet per acre. The woodlands of Allegany County in 1900 were almost without exception young. It was principally a deciduous forest, with only scattered remnants of young pine seedlings, descendants of mature pine trees that were once abundant. These mature pine trees, serving as the source for new pine tree seedlings, had escaped the axe in the first cutting but were of poor quality.

The older folks in the area told Sudworth that hemlock and chestnut oak were also once much more abundant. These species supported a thriving tanbark industry. One of the tanbark factories was located at Gilpintown near Flintstone. The bark peelers in the tanbark operation stripped bark from hemlock and chestnut oak, leaving the rest of the log rotting in the woods. The Gilpintown tanbark operation stopped operations many years before Sudworth's inventory in 1900 after depleting the area of needed bark.

Sudworth noted the rich tree species diversity within the forests of Allegany County, writing:

> *The peculiar position of western Maryland, intermediate between the North and the South, gives Allegany County a forest flora rich in species. The higher summits, coves and valleys exhibit a climate and soils closely similar to those of the more northern states, while the climate and soils of the lower valleys, glades and hills are characteristic also of the adjacent southern states. As a result, there is a conspicuous association of northern and southern tree species. This association is of more than passing interest, since the kinds of trees represented are of economic importance. Conifers and hardwoods of the middle South and North mingle here almost on the same ground.*

Today, one can observe the intermixing of trees from the South and the North at GRSF where the trees of the South (short-leaf pine and persimmon) intermingle with trees of the North (northern red oak and black cherry).

Sudworth listed seventy-two native trees he found growing in Allegany County, noting seven types of conifers and sixty-five hardwoods. On a map of Allegany County, he used a uniform green color to shade a single forest community type and called it "Woodland Areas." This map does not indicate the great diversity of the different forest communities mentioned in the narrative of Sudworth's report. The green shading on Sudworth's 1900 forest cover map shows about 70 to 75 percent of Allegany County covered by forestland.

If the forests of the region were to recover, the frequent forest fires that destroyed existing natural forest regeneration had to be controlled. These fires kept forest productivity at a standstill. If wildfires were not controlled, Sudworth grimly reported, "the present timber producing stock would eventually be exhausted."

In addition, if the forests were to return to health, the practice of allowing grazing in the woodlands had to stop. Livestock tramped, bruised and destroyed tree seedlings, compacted and damaged soils and exposed open areas to erosion.

Sudworth observed that despite the abuse that had occurred over many years, the forests continued to exist, "evidence of the greatest natural persistence in reproduction, which often takes place under very unfavorable conditions." Incredibly, Sudworth noted, there was no evidence that any species had been lost. Yes, there was an absence of large-sized trees such as white pine, white oak, short-leaf pine and hemlock, but forest regeneration was quite good. In 1900, Sudworth speculated that the forest would require one to two hundred years to fully recover and produce the supply of large white and short-leaf pine that the pioneers had found when they first settled in Allegany County.

# FRED W. BESLEY: MARYLAND'S PIONEERING FORESTER

*I heard Col. Greeley* [chief of the U.S. Forest Service from 1920 to 1928] *say on more than one occasion that Besley was one of the most effective of our early State Foresters. Col. Greeley said that if there had been more foresters like Besley, the forestry program would have advanced much faster than it did in our Country.*
—*W.D. Hagenstein, professional forester who worked under Colonel Greeley*

C an you imagine Maryland without a state forest or state park? Well, a little less than one hundred years ago, in 1905, this was the case. Green Ridge as a state forest did not exist. That all changed in 1906, when two brothers, John and Robert Garrett, made a generous donation of 1,917 acres of forestland in Garrett County to the State of Maryland. This tract of land is known today as Garrett State Forest, Maryland's first state forest. This donation, along with the passing of the 1906 Forestry Conservation Act, marks the beginning of the forestry conservation movement in Maryland. (Note: the legislature passed the law on March 31, 1906, and Governor Edwin Warfield signed it on April 5, 1906.)

Just consider one accomplishment of many resulting from the Garrett brothers' initial benevolence: there were zero acres of state public land in Maryland before their donation in 1906; today, there are fewer than 500,000 acres of state public land, making up about 10 percent of Maryland's land base. At the same time, the forestland base throughout the state increased from a little less than 30 percent one hundred years ago to 41 percent today. In 2002, more than 11 million people visited Maryland's state parks and state forests to enjoy all aspects of outdoor recreation. Today, we have state parks and state forests that did not exist one hundred years ago. Overall, in Maryland, there are sixty-six state parks and nine state forests. That's incredible, especially when you recognize that during the same time the population of Maryland tripled in size, from about 1.8 million in 1906 to about 5.6 million people in 2010. The forest conservation leaders of the past proved that it is possible to have economic growth while at the same time improving the forest resource base and quality of life issues. Somehow, they figured out that delicate balance.

With the donation of forestland, the Garrett brothers imposed several conditions that were soon legislated into law in the 1906 Forestry Conservation Act. The State of Maryland was to make

Fred W. Besley, circa 1906. *Courtesy Peggy and Don Weller.*

"adequate condition for its [forest] care." Maryland was required to establish a "State Board of Forestry" for the purpose of overseeing management of this land. The law stated that any additional gifts of land should be administered "as State Forest Reserves…to be used…to demonstrate the practical utility of timber culture and as a breeding place for game."

The 1906 Forestry Conservation Act addressed a variety of additional environmental concerns of that time: overcutting of timber, livestock grazing in woodlots and wildfires. All of these activities had a negative impact on forest regeneration. The law also provided guidelines for purchasing additional public lands. For example, the state could spend no more than five dollars per acre when purchasing additional public lands.

The 1906 Forestry Conservation Act mentions "parks" along with "forest reserves"—the two coincide right from the very beginning. The fact that no state parks existed in 1906 shows that this document was indeed progressive, forward-looking and optimistic. In 1912, Patapsco Forest Reserve near Baltimore became the first state park in Maryland.

The law stated that the State Board of Forestry was charged with appointing the state forester. This person was not to be paid more than $2,000 annually. More importantly, the state forester was required to "have practical knowledge of forestry." The person holding this position would not be just another political appointee but would be required to have professional forest management expertise.

The law, in essence, declared war on the "Age of Forest Exploitation." The law made it clear that the science of forestry would be the tool to heal Maryland's devastated landscape. The 1906 Forestry Conservation Act was so progressive and pioneering that it quickly put Maryland at the forefront of the national forestry conservation movement at the state level. The law called for the state forester to be in charge of carrying out its mandates in Maryland.

Besley's arrival in 1906 as Maryland's first state forester marks the beginning of a new forestry conservation era called the Custodial Period, lasting generally between the years 1906 and 1942. During this time, management placed its emphasis on protecting, nurturing and restoring the forest back to health. Foresters surveyed and mapped the forests for the first time. Foresters established tree nurseries in Maryland to aid in ecological restoration and acquired land for state public use. The Custodial Period coincided with the entire span of Fred W. Besley's career as state forester. During this time, Besley described his mission as reversing "destructive agencies, which for 150 years have been operating in the forests. Chief among them are forest fires, destructive cutting practices, excessive grazing, and the ravages of insects and fungus diseases."

Gifford Pinchot, the founding father of forestry in America and advisor to President Theodore Roosevelt, personally handpicked and oversaw Fred Besley's career with the U.S. Forest Service. Besley wrote about this in his unpublished autobiography. Besley mentions giving a talk at the annual meeting of the Colorado Forestry Association in Denver, where he was the principal speaker. Besley wrote, "I did not know until after the meeting that Gifford Pinchot, the Chief Forester, was in the audience. He had come in unexpectedly. I had a short talk with him at the close of the program and suspect this chance meeting with him had something to do with my selection for the Maryland job, although it was not even intimated then."

Besley mentioned that while he was working at Pike's Peak National Forest, he received an offer for the newly created Maryland state forester position. Besley wrote:

> *The offer came in a telegram from Gifford Pinchot and was delivered on horseback at our remote camp 10 miles from the nearest telegraph station and difficult to locate,—adding to our surprise. The offer came from Mr. Pinchot, Chief Forester, who had been asked to recommend a qualified man and who was guaranteeing a part of the salary. It appears I was selected because I had taken academic work at the Maryland State College and because of my good record at the Yale Forest School and later field work in Nebraska and Colorado known to Mr. Pinchot.*

In 1892, Besley graduated from the Maryland Department of Agriculture at the University of Maryland. He was first a teacher before he was a forester. He taught in a one-room schoolhouse for eight years in Fairfax County, Virginia (1892–1900). Then Besley worked for six years with the U.S. Forest Service as a "student assistant" under Gifford Pinchot (1900–6). During this time, Besley attended Yale Forestry School and graduated with honors (1904). Indeed, in 1906, it would have been hard to find someone more qualified than Fred Besley to be state forester. Besley's first day at work as a state forester was on June 20, 1906. His starting salary was $1,500, of which $300 was paid by the U.S. Forest Service.

In his role as a state forester, Besley set a very high standard for those who followed after him, pioneering and establishing many scientific forestry practices in Maryland. He was a trailblazing forestry pioneer. His life was one that truly inspires. When one looks at Fred W. Besley's career, one looks at the very beginning and advancement of forest conservation in Maryland, for they both coincide.

The idea to name this chapter "Fred W. Besley: Maryland's Pioneering Forester" came from the first conversation I had over the telephone with Helen Besley Overington, the daughter of Fred W. Besley. Helen told me, "Being one of the first state foresters, Father had to pioneer everything." In fact, Besley was the third state forester in the country. Pennsylvania was the first to have a state forester, in 1896; Wisconsin followed in 1904. Fred Besley wrote in his unpublished autobiography, "Maryland was one of the first states to select a technically trained forester to head up and direct all forest work. I[t] was real pioneering in Maryland, with no precedents or guides to follow."

When you think of a "pioneer" you also think of a "frontier." One hundred years ago, forestry was the frontier of science. Besley was Maryland's original trailblazer, bringing the science of forestry to Maryland. When Besley began his career as a state forester, the condition of Maryland's woodlands, in Besley's words, were "devastated." The seemingly "inexhaustible" timber resources were exhausted, consisting of cutover landscapes and seedling/sapling-sized forests. There was concern spreading around the country that America was running out of timber.

Besley called this period the "Age of Forest Exploitation." Timber volume greatly exceeded growth. Only 20 percent of forest cover

Baltimore Pike in eastern Allegany County, near Pratt Hollow. Taken from roof of present-day Knot Hole, circa 1920s. *Courtesy Ron and Gloria Jones.*

remained east of the Mississippi River. (Note: Maryland fared just a little bit better, with less than 30 percent of its original 95 percent forest cover remaining.) In the country, timber removal peaked in 1909 at 44½ billion board feet. Besley later used this benchmark number as a warning and red flag when annual harvests during World War II again began to approach 1909 peak harvesting levels.

Besley held traits in common with other contemporary conservation leaders like Gifford Pinchot and Theodore Roosevelt—they were sportsmen; they were avid hunters who shared a passion for wildlife, forests and wilderness. Besley especially enjoyed hunting waterfowl. Sportsmen like him, interacting closely with the land, would be among the first to notice any environmental problems.

In 1898, Besley met Gifford Pinchot at the U.S. Department of Agriculture in Washington, D.C. They were introduced through a member of the Sherman family. The Shermans were neighbors of the Besleys in Vienna, Virginia. Pinchot was only a couple of years older than Besley, and they took to each other immediately. The meeting changed Besley's life. Pinchot told Besley he ought to go into forestry. Besley later recalled this defining moment: "Pinchot was so boiling over with enthusiasm about forestry that then and there I adopted forestry as my career." Besley now wanted to be a forester!

However, it wouldn't be until two years after their first meeting, in 1900, that Pinchot secured enough federal funds to hire "student assistants" to work for him. Besley was one of 61 applicants chosen from 232 applications received. Besley could put on his résumé that he was personally handpicked by Gifford Pinchot for federal employment as a forester; that he was one of Pinchot's first field foresters; and that he trained and worked under Pinchot's direction for six years, between 1900 and 1906, with the U.S. Forest Service in nine different states.

Under Pinchot, Besley learned all aspects of forest resource management, both as a field forester and academically as a student at the Yale School of Forestry, where in 1904 he graduated with honors. At this time, Besley was thirty-two years old, married and the father of two children. During his time with the U.S. Forest Service, Besley supported himself and his family earning twenty-five dollars per month as a field forester and, in the winter, forty dollars per month transforming the collected field data into statistical forestry reports.

Between 1901 and 1902, during winter months on Thursday nights, Pinchot would invite his student assistants to his home in Washington, D.C., to listen to inspiring talks about forest conservation from various leaders in the field. These meetings became known as the "Baked Apple Club," for after a speaker finished his presentation, Mrs. Pinchot would promptly serve the students baked apples and gingerbread.

Can you imagine a president of the United States, one who is enthusiastic about forest conservation, attending one of your forestry meetings? Fred Besley had that experience! Helen Besley Overington told me: "They could have all the gingerbread and baked apples they wanted. One night Pinchot told them they were going to have a special guest...that special guest was President Theodore Roosevelt!"

As Maryland state forester, Besley pioneered methods to fight and detect forest fires. In Besley's words, forest fires "impoverished soil, destroyed reproduction, and [caused] damage [to] the large trees." In the 1920s, fire control was prominent in Besley's mind. In addition, he wrote in a report that year that one important function of his department was "to organize and maintain a state-wide forest fire protection program for 2,200,000 acres of forest land in the State." His program was very effective, reducing the average size of a forest fire from 203 acres in 1920 to 17 acres in 1927. In 1916, only three fire towers stood, all in Garrett County. The department built the Town Hill fire tower at Green Ridge in 1931. By 1942, during Fred Besley's tenure as state forester, forty-two fire towers had been constructed.

Besley left behind a rich legacy of documentation in written form and photographic images, many preserved as lantern slides, including glass lantern slides of Green Ridge. The Hall of Records in Annapolis, Maryland, archives these images.

Besley pioneered procedures and ways to gain public support to acquire lands for state forests and state parks in Maryland. It is an incredible accomplishment, considering that Maryland went from zero acres of state public land before 1905 to nearly half a million acres of state public land in 2006. When Besley retired in 1942, there were about 100,000 acres of state public land. In 1942, Green Ridge consisted of about 10,000. The groundwork was established and in place for acquisition of additional lands for the state forest and park system to grow; today, these lands benefit millions of people who visit Maryland's state forests and parks.

Besley pioneered state rights for management of state public lands. Green Ridge could have been a national park or part of the U.S. Forest Service under someone else's leadership. Besley believed that state government could better manage public lands and public parks than the federal government. Helen Besley Overington stated, "Governor Ritchie was very much for state rights, as was my father; Ritchie thought the closer to the people you got, the better it was, and so did my father." In 1927, Besley helped gain public support for legislation, giving the state primary rights to manage public lands over the federal government. This may be one of the reasons there are no national forests in Maryland today.

Besley pioneered the concept of linking scientific forestry with outdoor recreation. Besley was always looking for ways to gain public support and funding for forestry conservation and acquisition of additional lands. He believed that promoting outdoor recreation on public lands was one way to gain this support so that the public might be more likely to support the state in its efforts to purchase additional lands to practice scientific forestry. Helen Besley Overington stated, "Father believed that forest should not only be conserved, but that they should be used. Father was very interested in getting the public to use the land. Father thought this would bring more public support for conservation." In the early 1900s, camping in the outdoors for fun was a rather new concept to urban residents from Baltimore. Many thought that the only people who camped in tents either were in the army or were suffering hard times. Besley and his family camped sometimes a month at a time along Cascade Falls at Patapsco State Park. The public read *Baltimore Sun* newspaper advertisements inviting them to visit the state park and learn about camping. When they showed up, there would be the Besley family, giving public demonstrations on how to set up camp and cook.

Just before Besley retired, he came to Green Ridge State Forest to sign an agreement with a pioneering outdoor recreation organization, the Mountain Club of Maryland. Besley oversaw the signing of the lease agreement allowing the club the use of the abandoned Paw Paw Civilian Conservation Corps camp. His daughter, Helen, was a member of the club. During his visit, he gave a talk to members of the Mountain Club of Maryland at their campfire. Judging by his accomplishments as a leader and speaker, Besley, like Pinchot, had a gift to inspire people and motivate them into positive action. Helen said, "Father was a great storyteller…he especially loved to tell 'Paul

Bunyan stories.'" A heading from a newspaper article states this about Besley: "His Stories Are as Tall as the Trees He Protects." A photograph exists showing Besley conducting a campfire program at Green Ridge State Forest in 1941, one year before his retirement. Besley is surrounded by campers in casual clothing, while he stands in the middle of them looking like he's just walked out of church wearing a white shirt and tie. The campers' faces, without exception, are focused on Besley; whatever story he is conveying has certainly captured the campers' complete attention.

Besley pioneered environmental education. His presentations included topics about roadside tree care, forest pests and diseases and dendrology (tree identification). Besley used a lantern slide projector with glass slides to enhance his presentations. His lantern slide programs did not occur just in a sterile classroom environment; he often took his programs on the road, presenting them in open fields near woodlands. Helen remembers being with her father at these lantern slide programs. She recalled, "Father would jack up the back end of the car and, to power up the lantern slide projector [for the light bulb], attach a belt from the axle to the projector [generator]." He hung a sheet on a tree and stood beside it in front of the audience, while in the back, his children switched the slides back and forth on the projector. Helen remembered, "People were crazy [enthusiastic] to see them." A lantern slide projector was recently found by one of the rangers in an obscure place of the old Green Ridge Civilian Conservation Corps garage, where it had been sitting undisturbed since the 1930s.

At the time of Besley's retirement, a new era of forestry was beginning, called the Sustained Yield Period. The Custodial Period was ending with Besley's departure. State forests and state parks had grown to 100,000 acres at the time of Besley's retirement, with Green Ridge State forest at 10,000 acres, then making up 10 percent of the state forest system. Times were changing. The last CCC camp in Maryland permanently closed several months after Besley's retirement. The United States was now fully engaged in World War II after the recent bombing of Pearl Harbor.

In 1956, the year of Maryland Forest Service's fiftieth anniversary, Governor McKeldin presented a signed certificate to Besley. The

Fred W. Besley,
circa 1940s.
*Courtesy Peggy and
Don Weller.*

certificate recognized that Besley had "pioneered the magnificent tasks of conservation and reforestation." Through Besley's leadership, Maryland forestry practices "became the models for other States and Commonwealths of the Nation." The governor also noted how Besley was "recognized as the spearhead and inspiration for the great accomplishments...from 1906 to 1942."

Remarkably, Robert Garrett, who back in 1906 donated the first lands to become state forest, was present at this anniversary celebration. This must have been quite an event! Governor McKeldin told Besley, then eighty-four years old, that there was still much work for him to do and accomplish, reminding him that Noah was three hundred years old before he started building the ark.

# FORESTRY: A PRESCRIPTION TO HEAL THE LAND

*All good forest management starts with good field data.*
—*Steve Resh, forestry professor at Allegany College of Maryland*

Modern-day forest inventories provide an "accurate representation of the flora and fauna within a forested ecosystem," John Jastrzembski, a forestry professor at Allegany College of Maryland, has stated. Forest inventories consist of useful and valuable forest data. Often, detailed maps identifying locations of the forest resources are prepared showing, for example, stands of trees with similar characteristics, such as species composition. Foresters map trees of similar species and age and group them together into stands.

From this data, foresters write detailed forest management plans, which are an essential management tool that foresters use to manage the forest. The forest data in the plan aids the forester in understanding the condition and health of the forest as well as useful information about tree species, their size, age and estimated volume in a designated area. Modern-day foresters also collect other information, such as understory vegetation and how much woody debris is present; however, back in the early twentieth century, foresters focused their inventories on trees.

One of Fred W. Besley's first tasks in 1906 when he assumed responsibilities as Maryland state forester was to conduct a statewide forest inventory. This monumental project involved inventorying in

Maryland "every tract of woodland five acres or more...sketched on a topographic base map, on a scale of one mile to the inch, and its general characteristics noted." Between 1906 and 1916, he accomplished this work with a staff numbering fewer than the fingers on his hand. Besley recalled, "I tramped every cow-path in Maryland making it." It was indeed a magnificent accomplishment!

Besley's book, *The Forests of Maryland*, published in 1916, summarized field information that he had collected over ten years, including the Allegany County Forest Inventory of 1908–9. It was a pioneering accomplishment, as it was the first statewide forest survey published in the country. Besley wrote that his book made available in "condensed form and orderly manner our present forest resources, their value to people of the State, and how these resources may be best conserved by wise use, not only to supply present needs, but anticipate the needs of the future." Besley believed that the forests were very important economically to Maryland citizens because "in their value as natural products, forest resources of Maryland ranked second only to agriculture."

Allegany County was one of the first counties Besley inventoried, conducting fieldwork in 1908–9. Perhaps he picked Allegany County as one of the first counties to inventory because at that time Allegany County, at 62 percent forest cover, had the "largest percentage of forestland of any County in the State." In addition, Besley had in his possession the 1900 Allegany County forest inventory that Sudworth had completed under Bullock Clark's direction as a base reference on which to build his inventory.

Before he published his book in 1916, Besley had published a pamphlet in 1912 called *The Forests of Allegany County*. Besley wrote that the primary purpose of the Allegany County forest survey was to determine "the extent, character, and condition of the woodlands, with the view of suggesting improvement."

Besley noted that although the county had 62 percent forest cover, "of this, not greater than one percent was virgin forest, the remaining ninety-nine percent having been cut over once, if not several times, since the settlement of the country." Besley wrote that then the "virgin woodland covered ninety-five percent of the county's total area."

The original character of the forest in Allegany County had greatly changed by the time Besley arrived on the scene in 1908. Besley observed that excessive timber cutting and wildfires had eliminated most of the trees of greatest timber value, such as short-leaf pine, hemlock and white pine. Loggers needed to implement new improved methods in timber removal practices.

Where Sudworth had mapped just one category, "woodlands," Besley divided and mapped the forested landscape into three categories: hardwood; pine; and mixed pine and hardwood. In 1912, Besley noted that the composition of the Allegany County forested landscape was 78 percent hardwood forest cover, 20 percent mixed hardwood and pine and 2 percent pine.

According to Besley, only a few acres in the county had a stumpage of five thousand board feet or more. His summarized that in 1908–9 there were 174 acres of mature hardwood, 6 acres of mature pine and 442 acres of mature mixed stands of pine and hardwood.

Like Sudworth in his 1900 forest inventory, Besley observed transitions in forest composition as one ascended the mountains from the bottom to the top. On the lower slope, he observed that dominant species were white oak, sugar maple, basswood, red oak, ash and, more sparingly, white pine and hemlock; on the upper slopes were chestnut, hickory, red and black oaks and Virginia pine; and on the ridge tops were chestnut oak, pitch and table mountain pine. Besley identified fifty-four distinct species of native trees growing in Allegany County, many of them of commercial importance. Today, five native pines grow at Green Ridge State Forest: white pine, Virginia pine (Besley called Virginia pine "scrub pine"), short-leaf pine and pitch pine. Besley's inventory only mentions four native pines, including all of the above with the exception of short-leaf pine, indicating that this species was generally absent from the Allegany County woodlands in the early part of the twentieth century.

Besley concluded that the forests of Allegany County were in very poor condition from a long history of overcutting and frequent forest fires. The state forester wrote that the character of the original forest had changed, noting that white pine and hemlock had made up a much larger percentage of the original forest; however, they

Baltimore Pike on the east side of Polish Mountain near Pratt Hollow, circa 1920s. Photograph by Fred W. Besley. *Courtesy Jenny Bond.*

were the first timber trees to be exploited due to their great value and ease of logging. After the loggers cut the conifers, hardwoods quickly took possession of the forest openings, making it difficult for conifer trees to reestablish themselves.

Wildfire often followed logging operations, greatly exasperating conifer reproduction as wildfires destroyed the thin-barked young pine seedlings. Besley wrote that conifers like white pine lacked "the power of regeneration by sprouts, have largely been replaced with more persistent hardwoods that possess the power of growing from the roots and stumps after the stems have been killed by fire."

The forests were unproductive, in Besley's words, "primarily due again to the two main culprits, abusive logging practices, and frequent uncontrolled forest fires." Besley noted that the practice of timber overcutting in 1908–9 was occurring at a 50 percent higher rate than annual growth. Besley warned to expect negative consequences if this abuse continued: "At the present rate of cutting the supply will last but twelve years, if the rate of growth and the forest area remain the same." In other words, Besley was warning that the timber supply in Allegany County would completely run out by 1920! Besley predicted that this could have disastrous consequences, for while demand for timber supplies by industry was increasing, timber supply was decreasing at a rapid rate. He predicted a timber famine if the forests continued to be plundered.

Besley also observed the extensive land clearing occurring in 1908–9 in eastern Allegany County. The Mertens Valley Orchard Company was clearing and burning off the land for "fruit-growing" in the area of present-day Green Ridge State Forest. This was of great concern to Besley, for the Mertens Valley Orchard Company owned more than thirty-five thousand acres. Therefore, the Mertens operation had the potential and capability to clear about one-fifth of the forested land base in Allegany County!

Besley proposed radical changes to restore, protect and manage the forest to improve its productivity. He recommended that wood products be used more conservatively and with greater economy than the current abuse and waste that was occurring. He believed that with the elimination of forest fires, the annual growth of the forest would double. In addition, with proper forest management, annual growth would quadruple by implementing improvement and thinning practices. Finally, Besley reemphasized that forest productivity would greatly improve by "eliminating destructive methods of harvesting" and by practicing conservative, economical and efficient uses of wood products. In essence, Besley was stating that forest conservation, when properly implemented, would restore and heal the land.

# THE VAGABONDS MAKE
# CAMPING POPULAR

*I like to get out in the woods and live close to nature. Every man does. It is in his blood. It is his feeble protest against civilization.*
*—Thomas Edison, 1921, while camping in western Maryland*

E arly in the afternoon of July 27, 1921, three of the most well-known men in America traveled westbound through the Green Ridge area along the Baltimore Pike in eastern Allegany County, Maryland. They were Thomas Edison, world-famous inventor; Henry Ford, automobile manufacturer; and Harvey Firestone, tire magnate. They were traveling that day to present-day Swallow Falls State Park in Garrett County, Maryland. Here they planned to camp several more days after camping nearly a week along Licking Creek, about six miles east of Hancock, Maryland.

Edison, Ford and Firestone were business partners. Their working relationship transformed into a bond of great friendship through the experiences they shared camping together each summer for about two weeks from 1915 through 1924. When on these camping trips, these wealthy captains of industry called themselves "vagabonds," for they took great pleasure in roughing it together in the world of nature, far away from civilization.

On this particular summer day, the vagabonds were in great spirits. The president of the United States, Warren G. Harding, had just visited and camped with them for a night at Licking Creek. The press corps that followed them on this trip documented their every move, writing newspaper articles, taking photographs and publishing their activities in newspapers across the country. The public eagerly read about their camping adventures and seemed to be interested in everything they did, from chopping firewood, horseback riding and fishing to sitting around campfires. They also took an especial interest in what time the campers turned in at night to sleep "under the canvas."

The loud sounds of the motorized caravan traveling along the Baltimore Pike in eastern Allegany County that July day broke the quiet of the rural countryside and certainly drew the attention of anyone within hearing distance. They would have seen Thomas Edison, the self-appointed navigator, leading the motorcade, sitting in an open touring car with compass and map in hand. Edison told reporters their plans that summer were "to get to the wildest sections" of the "Cumberland Mountains."

As the caravan of automobiles crested Town Hill, perhaps they took time to stop and see the spectacular scenery from the overlook. Popular postcards at the time described this area as the "beauty spot of Maryland." The vagabonds certainly would have noted the newly constructed Town Hill Hotel, said to be the first "motel" constructed along the Baltimore Pike in western Maryland that catered to automobile traffic. Today, the Town Hill Hotel is still in business and is a popular tourist destination, operating as a bed-and-breakfast.

On Green Ridge Mountain, the vagabonds would have observed as far as the eye could see a seemingly unending apple orchard stretching as far west as Polish Mountain. Most of the apple trees they saw were part of the Mertens Green Ridge Valley Orchard Company operation, which had gone bankrupt three years earlier in 1918. Today, much of this area is Green Ridge State Forest; however, in 1921, this land was still in private holdings.

In fact, in 1921, little more than 2,746 acres of state-owned "forest reservations" existed. These public lands were located at Skipnish Reserve in Garrett County (888 acres); Swallow Falls Reserve in Garrett County (823 acres); Kindness Reserve in Garrett County (206 acres); and Patapsco Reserve in Baltimore and Howard Counties (829 acres). As stated, Green Ridge State Forest did not yet exist.

The national publicity that followed these celebrated men on their summer adventures introduced the public to the pleasure of motorized recreational touring, outdoor recreation and camping. Historians record that the vagabonds' camping trips were the "first notable linking of the automobile and outdoor recreation. Edison's comments in western Maryland captured the public imagination across America: "The woods will get you if you don't watch out…Stay out close to nature and you won't want to come back to the civilizing influences of trolley cars, telephones, porcelain bathtubs and nickel plumbing."

In the early 1900s, camping for fun in the outdoors was a relatively new concept. At that time, most urban dwellers believed that the only people who camped in tents either were in the army or were suffering hard times. Many Americans worked a six-day week, laboring ten or more hours each day. There was little time for recreation. As more leisure time became possible in the 1920s, people desired to tour in their automobiles and experience for themselves vagabond-like adventures.

The vagabonds' well-publicized adventures caused a sensation and stirred a national movement for motor touring and camping. The thing to do was "to buy a Ford car and then some camping equipment and see America first." The vagabonds popularized the idea of camping on wheels in faraway places, creating a demand for the acquisition of additional land for state forests and parks where the public too could camp and recreate in the outdoors. This public demand planted additional seeds of opportunity, making it possible ten years after the vagabonds' visit for the establishment of Green Ridge State Forest in 1931.

From 1920 through 1930, Americans started to become more affluent and, at the same time, gain more leisure time. For the first time, many families could afford to purchase automobiles. Like the vagabonds, they were looking for places to camp and experience the great outdoors. Tourists and campers needed more public forests and parklands to meet their needs. To meet the growing demand for more public recreational land, Stephen Ting Mather, the first director of the National Park Service, encouraged natural resource state agencies across the country to step it up and acquire more state forests and parks to supplement the growing system of national parks. Fred W. Besley took Mather's suggestion to heart and intensified his efforts to expand public lands in Maryland for state forests and state parks.

# THE FOUNDING OF
# A STATE FOREST

*A forest is God's garden.*
—*John Mash, forest manager of Green Ridge State Forest (1971–88)*

The fact that Green Ridge State Forest exists as public land and not a suburban neighborhood full of houses and shopping areas is a modern-day miracle. It was a fortuitous twist of fate that GRSF is not a suburban housing development with retail stores and shopping malls, something like the town of LaVale, Maryland, with a population of five thousand people or more; instead, destiny intervened at Green Ridge when Fred W. Besley entered the scene in the late 1920s.

In the early 1900s, the Mertens family, who then owned the land, managed this area as a colossal apple orchard enterprise. The Mertenses subdivided the land and sold more than 3,500 ten-acre lots to individuals across the country. In 1918, the Mertenses went into bankruptcy, and with the failed orchard business, the idea of a utopian society living off the fruit of the land died with it.

The Green Ridge area was on Besley's radar as a possible place to establish a state forest for a long time. He was familiar with the area. After all, he had conducted a forest inventory in Allegany County, which included Green Ridge, in 1908–9. However, twenty-three years passed before Green Ridge became a state forest in 1931.

In the late 1920s, Besley began focusing his attention on eastern Allegany County as a place to establish another state forest. From his office in the Fidelity Building, room #1411, off Charles Street in Baltimore, Besley directed his western Maryland district forester, H.C. Buckingham, to keep an eye on additional opportunities to purchase land for the new state forest.

Allegany County citizens should celebrate August 7, 1931, with an annual county public lands festival. It was on this day that the Grove family sold to the State of Maryland a 1,735-acre tract of land in eastern Allegany County that established Green Ridge State Forest. From these humble beginnings, Green Ridge State Forest became destined to become the largest contiguous block of forestland within the Chesapeake Bay watershed in Maryland, enjoyed by millions of outdoor enthusiasts over the years.

At the very beginning, people called the new state forest "Belle Grove State Forest," as the first tract of land purchased by the state was located in the Belle Grove area of eastern Allegany County. Several months later, James Price sold an additional nearby 394¼-acre tract to the state. The Price tract, located immediately on the north side of Belle Grove State Forest, made the new public land a little over 2,000 acres.

H.C. Buckingham, western Maryland district forest supervisor, oversaw the first work accomplished on the new state forest, a project which involved surveying and marking the state boundary line. The February 1932 *Maryland Forest Warden Newsletter* documented those individuals who carried out this assignment: "Forest Wardens Frank Davis, Urner Wigfield, Eugene P. Sipes, and Watchman Clem Wigfield have been engaged in surveying a portion of the Belle Grove State Forest under the direction of Henry D. Schaidt of Cumberland." From these humble beginnings, the work of building a great state forest began.

With the acquisition of the Belle Grove tract, Besley focused on other opportunities to purchase additional land in eastern Allegany County. Besley's vision of establishing a large state forest in eastern Allegany County really seemed to be within reach when he learned from Buckingham of a tax sale involving a large tract of land, over eleven thousand acres, owned by the Allegany Orchard Company. In a letter dated September 26, 1931, Besley wrote Buckingham: "I believe there is a good opportunity for us to get a large block of forest land in eastern Allegany County, suited for a state forest, and concerning which there would be little question as to its value for other uses."

Besley further instructed Buckingham to pursue the purchase of any additional desirable areas for two dollars an acre or less. The state forester understood that they might have to pay an additional amount per acre for more sought-after tracts of land, "for which the value by reason of the timber, location, or some other factor makes it worth much more." In addition, Besley advised Buckingham to be forward-looking and vigilant to see how much land "can be blocked in and purchased at the present time, or in the next few months, at a reasonable price, and where the title can be vouched for." One can sense Besley's sense of urgency in 1931, as if the window of opportunity to purchase additional lands would soon close. Besley cautioned Buckingham, "We will have to act without delay and with as little noise as possible."

In August 1932, the successful acquisition of the Allegany Orchard Company and Carpenter tract, along with the additional size increase in acres, caused the name of the state forest to change. The August 1932 *Maryland Forest Warden Newsletter* announced the following:

> The name of Belle Grove State Forest has been changed and is to be known as the Green Ridge State Forest. Two large tracts have been purchased there. The Carpenter tract of 1,994 acres is situated between the National Pike and the Pennsylvania State Line, on the east side of Polish Mountain. The other area is part of the Allegany Orchards Corporation property, and consists of 11,777 acres. The location of this tract is adjacent to the Carpenter tract and extends from the Pike to the Potomac River between Town Hill and Town Creek. The area does not actually touch the Belle Grove area but it is very close and will be handled as one forest.

The acquisitions of Belle Grove, the Price tract, Allegany Orchard Company and the Carpenter tract became the chief cornerstone properties that built the Green Ridge State Forest we know today.

Besley oversaw an explosion of state forest acquisitions across the state. By the end of 1932, the state had significantly increased the size of the state forest system to 49,073 acres: Green Ridge (16,177 acres); Swallow Falls (4,596 acres); Potomac (6,073 acres); Savage River (16,329 acres); Fort Frederick (189 acres); Patapsco (1,116 acres); Cedarville (2,631 acres); Doncaster (1,157 acres); Seth (65 acres); and Pocomoke (740 acres).

The men who worked at Green Ridge State Forest in the fall of 1932 prepared for fall fire season. Forest Warden George Wigfield reported the following: "I have had the pleasure of visiting the Green Ridge State Forest this fall and found a wonderful piece of work being accomplished. Under the supervision of District Forester Buckingham, the cutting of fire lines, opening of trails, and old springs on the property have gone far towards preventing fires."

Eugene Sipes, first resident warden at Green Ridge State Forest. *Courtesy Department of Natural Resources.*

Eugene Sipes, Green Ridge State Forest's first resident forest warden, is the prototype of a long line of hardworking state forest employees that followed him over the years to the present. In the earliest records of GRSF, especially in 1932 and 1933, one year before the Civilian Conservation Corps camps were established, Sipes is recorded doing a little of everything in forest conservation work: staffing and overseeing operations of the Town Hill fire tower; laying out growth study plots on thirty acres involving releasing young white pine trees established in the understory by thinning and removing dead overstory trees; establishing the first forest plantation in the spring of 1933 by planting fifty-four thousand red pine trees and, at another site, for experimental purposes, planting two hundred loblolly pines; implementing the first forest product sale involving the sale of pine pulpwood severely attacked by the southern pine bark beetle; and conducting the first environmental education program at Green Ridge with a forest conservation talk at the Belle Grove campsite, which "aroused much favorable comment."

# Roots of the National Big Tree Champion Contest Planted in Maryland

*The best time to plant a tree was twenty years ago—the second best time is today.*
*—Fred W. Besley*

The Allegany County Forestry Board annually sponsors a Big Tree Champion Contest. Presently, there are more than thirty-five big tree champions, each representing the largest tree species of its kind growing in Allegany County, Maryland. However, there are many more tree species in Allegany County still without a big tree champion. The board regularly encourages citizens around the county to submit nominations for big trees. Members of the board go out in the field and measure newly nominated trees to confirm the species and their size. Yearly, on Arbor Day in April, the board announces the biggest tree for each nominated species.

The roots of the Big Tree Contest were planted in Maryland. Fred W. Besley, Maryland's first state forester, is the father of the National Big Tree Champion Contest. In his booklet, *Big Tree Champions of Maryland: A Record of the Largest Trees of the Principal Species*, printed in 1956, Besley discussed the birth of the Big Tree Contest. He wrote that the forests of Maryland were rich in different kinds of trees, "probably more than 250 native tree species, and there was a universal interest in notable trees." Because of this, Besley was inspired "to organize a Big Tree program in which would be collected measurements and photographs of the distinguished trees of Maryland." It appears that until 1925, Besley worked alone on this project.

By 1925, the interest in large trees and noted trees had so increased, Besley wrote, "that the Maryland Forestry Association sponsored a state-wide Big Tree Contest. Prizes were offered, rules were adopted, and wide publicity given to secure as many entries as possible. Each tree species was classified separately so that species like dogwood and persimmon would not have to compete with such larger trees as oaks and elms."

Besley developed the method of measuring big trees that was adopted by the American Forestry Association, with only some slight modifications. Besley wrote about this: "At this time [early 1900s] there were no standard measurements of trees, so it was necessary to draw up standards to insure fair comparisons. The author devised the following standards. To qualify as a tree, the specimen must have a single stem or trunk for at least $4\frac{1}{2}$ feet above the ground level and a total height of 15 feet." Besley's method took three important measurements involving the trunk circumference, crown spread and height of a tree.

In the first statewide Maryland Big Tree Champion Contest of 1925, Besley notes that 450 entries were received. Besley acted as "umpire in measuring those [trees that] appeared to be competitors in the prize winning class." After all was said and done, the first Maryland Big Tree Champion list contained 155 species, among them the Wye Oak, the largest white oak ever recorded. In 1937, this list was revised and published.

We learn in the book *Wye Oak: The History of a Great Tree*, by Dickson J. Preston, that Fred W. Besley was behind the efforts to expand the Maryland Big Tree Contest to a national level. Preston writes the following:

> *In 1940, he* [Besley] *suggested to the American Forestry Association (of which he was by now a senior member of the board) a means of putting the Wye Oak and other declared* [Maryland] *national champions to the test. His proposal was a national contest along the lines he had been conducting in Maryland. Readers of* American Forests...*would be invited*

Liberty Tree at St. John's College, Annapolis, Maryland. It held the title as the largest yellow poplar in the United States. It was destroyed by Hurricane Floyd in 1999. A seedling from the tree is now planted at Oldtown, Maryland. Photograph by Fred W. Besley, circa 1920s. *State Archives.*

*to send in measurements of trees they thought should be national champions, and the winner would be chosen by the Besley system of measurement.*

Therefore, when you locate and nominate a Big Tree candidate, you are carrying on a legacy started here in Maryland by Fred W. Besley. Indeed, it is a noble cause, for the purpose of the Allegany County Big Tree contest is the same as announced for the first National Big Tree Champion contest held in 1940: "To locate 'the largest living specimens of American trees and focus attention on the benefits of conserving these cherished landmarks.'"

# THE TOWN HILL SENTINEL QUIETLY WATCHES THE FOREST GROW

*Forest watchers keep a high, lonely vigil…the tower is 110-feet tall and bright silvery paint on its steel angle-iron frame glitters in the afternoon sunlight above the 2,135-foot peak of Warrior Mountain south of Flintstone…at the very top is an eight foot square wood and glass cab which sways gently with the autumn breeze…visitors can enter through the trap door of the floor enclosure…the scenery is pretty much as the Indians would have found it if they had built towers…only a few roads and farms are visible to show that someone lives in Allegany County.*
*—Bernard E. Sitter,* Cumberland Times-News, *1960s*

The Town Hill forestry lookout tower is a well-known landmark in eastern Allegany County at Green Ridge State Forest. The tower was constructed in 1931, the same year GRSF was established. The tower overlooks the state's first acquisitions of about two thousand acres around the Belle Grove area. The Town Hill tower was the third fire lookout tower built by the Maryland Forest Service in Allegany County, following the construction of Dan's Rock fire tower in 1921 and Warrior Mountain fire tower in 1922.

Since the very beginning of the Maryland Forest Service in 1906, protecting the forests from wildfires has always been a very high priority. Maryland's first state forester, Fred W. Besley, developed the state's earliest systematic program for fighting forest fires, and fire towers played an important role in his plans.

About 1910, Besley began commissioning an elite corps of volunteers called forest wardens to fight forest fires. By 1935, the forest warden roster had grown to about 650 individuals. Their responsibilities included coordinating "registered crews" to fight and extinguish forest fires, post fire warning notices and enforce forestry laws. Beginning about 1915, the Maryland Forest Service developed a plan to build thirty to thirty-five forest lookout towers throughout the state to help improve the forest wardens' effectiveness and response time to forest fires. In 1915, the Maryland Forest Service built the first forestry lookout tower in the state on Meadow Mountain in Garrett County.

The first generation of forestry fire towers, like that at Meadow Mountain, were little more than fancy wooden treehouses sitting on top of four long metal legs anchored by guy wires. Between 1915 and 1927, eleven towers were constructed; by 1937, thirty fire towers existed; and by 1940, thirty-two fire towers were in use throughout the state.

By the 1920s, the Maryland Forest Service started purchasing a prefabricated fire tower designed and made by Aermotor, located then at Broken Arrow, Oklahoma. The tower was designed to be shipped in parts and assembled on site. Towers ordered for the East Coast were first shipped by truck to Chicago before reaching their final destination. In 1930, an Aermotor fire tower cost from $400 to $800, depending on the height of the tower. Today, a similar tower would cost ten times that much. For example, a fire tower built in 1979 in our neighboring state of Pennsylvania cost more than $100,000.

The height of each fire tower varied based on its location, the primary consideration to give the lookout observer an optimum 360-degree view of the surrounding forested landscape, often more than twenty miles in all directions. The Town Hill fire tower is 80 feet tall and sits atop Town Hill 1,685 feet above sea level, the cabin on top peeking just above the forest canopy, with minimal impact to the overall aesthetics of the forest.

During the spring and fall fire seasons, forest wardens scheduled and assigned forest fire tower lookout observers. During times of moderate fire danger, the towers were generally occupied between

Aermotor fire tower like the one at Town Hill. This one is in Burtonsville, Maryland, circa early 1940s.
*Courtesy Jenny Bond.*

10:00 a.m. and sunset. With binoculars, the lookout scanned the 360-degree horizon every fifteen minutes. Upon observing a smoke or fire, the observer would identify the general location of the fire with a piece of equipment called the "Osborne fire-finder," sometimes called an alidade, which gave an azimuth compass reading in degrees from 0 to 360 from the tower to the fire. After sighting the smoke through the alidade peephole, much like lining up a target using a rifle scope, the lookout obtained an azimuth reading to the fire and reported this information immediately by radio or phone to the forest warden.

A second azimuth reading was then obtained from another nearby tower; in the case of Town Hill, this information often came from the lookout observer at Warrior Mountain fire tower. By triangulating

Pete Bond sighting smoke with alidade in Long Hill fire tower in the early 1940s. *Courtesy Jenny Bond.*

the two azimuth readings, the forest warden determined the exact location of the fire. A "smoke chaser" was then dispatched to the fire to obtain additional information needed for fighting the fire.

The introduction of fire towers established an early romantic stereotype of the forest service and of the work of the forest warden, giving a completely new outlook to the job. The people who spent many hours in the Town Hill fire tower—individuals like Wade Mallow, Amelia Dixon, Vivian Roberts and Jack Hensley—were true pioneers in their own right, contributing to Maryland's early forest fire control efforts and helping to save lives and valuable natural resources.

Today, the Town Hill fire tower stands as a quiet sentinel that has overseen the growth of Green Ridge State Forest over the past seventy-seven years, from 2,000 acres in 1931 to 43,560 acres in 2010. It is a monument to Maryland's early forest conservation and forest fire control efforts.

# A Tree Army Comes to Green Ridge

*The CCC's was a great conservation movement...conservation of our natural resources, but also conservation of our human resources. Those young men had the experience of a lifetime.*
—*Joe Davis, CCC-era professional forester hired by State Forester Fred W. Besley*

In the early 1930s, during the Great Depression, millions of men were out of work and could not find jobs. The then-president of the United States, Franklin Delano Roosevelt, addressed this crisis with the following encouraging words when he accepted the presidential nomination: "Let us use common business sense...We know that the...means of relief, both for the unemployed and for agriculture, will come in a wide plan for the converting of many millions of acres of marginal land and unused land into timberland for reforestation...in doing so employment can be given to a million men." When he spoke these words, the president-elect had his eye on rehabilitating the devastated forests of the United States on both private and public land. This work would give full-time, meaningful work for millions of the unemployed.

True to his word, just seventeen days after his inauguration, Roosevelt gave a message to Congress proposing legislation "to help relieve distress" and to "build men" with a program promoting conservation-related work on federal, state and privately owned forestland. Ten days later, Congress enacted legislation authorizing

"President Roosevelt's Emergency Conservation Work Program," better known a year later as the Civilian Conservation Corps (CCC).

The CCC, from 1933 to 1942, was a key to the greatest national public forest and park development effort in American history. During this time, the CCC employed more than 3 million young men between the ages of eighteen and twenty-five in conservation work across the nation. In Maryland, under Fred W. Besley's direction, Roosevelt's "tree army" of 35,800 men built and improved its state forests and parks. The CCC conducted lasting conservation work on both public and private lands by constructing roads, buildings and pavilions; erecting fire towers; fighting forest fires; planting millions of trees; and stabilizing soil erosion.

The essential purpose of the program was to restore and build confidence in America's young men shaken up by unemployment caused by the hardships of the Great Depression. This worthwhile, productive work on the nation's forests would "help protect, develop, and perpetuate existing forests; help to prevent soil erosion, which aggravate damage from floods; and help to establish new forests, and reestablish old forests." A 1933 government pamphlet promoting the CCC stated, "The jobs are important; the building of men is also important. Fortunately the two go hand in hand: The men need the forests, and the forests need the men."

Many of the young men were processed Camp Holabird near Baltimore, Maryland, an army motor depot. Personnel lived in permanent brick and wooden barracks. The CCC induction center was off to one side. It processed about one hundred enrollees at a time, and during the week it took for enrollment the new enrollees lived in pyramidal tents, four bunks to a tent. The CCC program was administered by the army, so it followed the army pattern in most things. After physical exams and interviews, enrollees drew clothing—the regulation army drab uniform (all heavy woolens), blue denim fatigue clothes and heavy hobnailed field shoes.

In a *Baltimore Sun* magazine article written in the 1970s, Francis Zumbrun Sr., an enrollee in a Maryland CCC Camp in 1934, said this about his experience:

> *I have never seen spirits run higher in any organization. We were learning something, doing something meaningful, and getting paid*

*for it. Our pay was $30 a month. A check for $25 was sent by
the paymaster to our homes, to be used by our parents. Small
as it may sound now, the monthly check, in many homes, made
the difference between getting enough to eat and missing a few
meals. The remaining $5.00 was personal spending money for
the month...I read somewhere that in his tour of duty, the average
C.C.C. enrollee gained eleven and 1/2 pounds and grew half
an inch. I went from 128 pounds to 160 pounds...With ax,
grubbing hoe and all the other pioneer tools...we thinned out the
runt trees, and cleaned up the forest floor to reduce fire hazard...
After a heavy breakfast, we piled into trucks and were on the
job working at 8 or 9 o'clock. A truck brought out coffee and a
mountain of sandwiches for our noon break. We knocked off
around 4 P.M. After a shower and some chow, we were on our
own for the rest of the night. Nights and weekends we could
wear civvies if we liked when we walked to a movie...or had
a date, but we were very proud of our uniforms and usually
wore them...We...wore Army clothing and got a taste of Army
regimentation and discipline. I don't know of any C.C.C. boy
who didn't become more of a man because of it. At its peak,
the C.C.C. had something over 2,500 camps across the country,
working more than half a million men. About 3,000,000 served
hitches in it from 1933 until the beginning of World War II put
an end to the program. Thousands of these men learned useful
skills—as cooks, bakers, warehouse men, mapmakers, truck
and tractor operators, radio operations, welders, quarrymen. At
the same time they put themselves into good physical shape and
developed personal discipline that stood them well for the rest of
their lives. This is to say nothing of the countless lakes, rivers,
and shorelines they cleaned up, the countless trees they planted, the
preservation work they did around the country's national shrines
and scenic spots.*

From 1933 to 1942, an average of twenty-one CCC camps
operated in Maryland at any one time. The State Board of Forestry
sponsored fifteen CCC camps. Three of them were located at Green
Ridge State Forest: Green Ridge Camp (S-53, 324, Flintstone,

Maryland); Town Hill Camp (S-58, 1359, Paw Paw, West Virginia); Little Orleans Camp (S-61, 377, Little Orleans, Maryland). Note that the first number of the camp was a state designation, the second number was a federal government designation and the name of the camp was the location of the closest post office to the camp.

Fred W. Besley served as the director of emergency conservation work to the CCC forest camps in the Third Corps Area of the War Department. His responsibilities were not only in Maryland but also included Virginia, the District of Columbia and Pennsylvania. One of Besley's responsibilities was to oversee the movement of men from military conditioning camps to the "National Forest Camps" in the Third Corps area. For example, the May 1933 *Forest Warden Newsletter* noted that Besley "approved the organization of twenty-one camps of about 200 men each on the Pennsylvania State Forest Areas."

Karl Pfeiffer, assistant state forester, did much of Besley's field legwork as his deputy. In early 1933, Pfeiffer, Buckingham and an army lieutenant colonel inspected several proposed forest campsite locations in western Maryland.

In the early spring of 1933, authorities announced the locations of the first group of Maryland's CCC camps. The first seven CCC sites were at Green Ridge State Forest along Fifteen Mile Creek; Potomac State Forest, Swallow Falls and New Germany in Garrett County; Camp Ritchie in Washington County; Fort Meade in Anne Arundel County; and Fort Washington in Prince George's County. Each two-hundred-man CCC camp was a self-contained community with cooks, barbers, machinists and other support personnel to sustain the operation. The CCC camps were set up, maintained and administered by the War Department. The State Board of Forestry was only responsible for directing the work activities of the men in the forests.

This was a busy time for Besley. Between June 27 and July 1, Besley personally visited all the CCC camps in western Maryland. He was very pleased with the progress the men were making in the field and personally witnessed a frenzy of productive activity. About a week later, on July 6 and July 7, he also visited several Pennsylvania CCC camps in his role as the representative conservation director of the Third Corps Area of the Army.

Henry C. Buckingham, Joseph F. Kaylor and Karl Pfeiffer. Kaylor succeeded Fred W. Besley as state forester, and Buckingham later succeeded Kaylor. Circa 1943. *Courtesy Jenny Bond.*

## GREEN RIDGE CCC (S-53) CAMP

The Green Ridge CCC camp was located along Fifteen Mile Creek just south of present-day Interstate 68, off exit 62. It officially opened on May 22, 1933. The first group of young men arrived from Fort Howard, Maryland. The *Forest Warden Journal* noted that since the camp opened, "these Baltimore boys have been introduced to many kinds of work to which they were more or less strangers, but many of them have taken hold of their new work with both hands and are doing a good job of it."

The new enrollees learned the proper and safe use of tools such as the pick, shovel, axe, hammer, drill, firefighting pump and rake, as well as carpentry tools. They did not have all the comforts of home when they first arrived, staying in tents provided by the army as the wooden barracks were still under construction, as were the mess hall and showers. Everyone looked forward to the completion of the construction of the shower building, tiring of taking a bath with bucket water. The CCC staff constructed a seven-foot fence around the garage compound where eight trucks could be parked and were in close access to the gasoline pump. Outside the garage area, staff constructed as a tool house and a blacksmith shop.

Officials of the CCC were looking for men who demonstrated "ability, energy, and patience in large portion" to ensure the success of the work done at camp. Meeting that description, four forest wardens were hired by the state: Eugene Sipes, GRSF's first resident warden; Cecil Ward, C.V. Crocks; and R.H. Wempe. The support staff at the camp included David W. Sowers, project superintendent; Mr. Wigfield, blacksmith, C.W. Shlagel, clerk; H.L. Stuart, engineer; and H.K. Cheney, mechanic.

Some of the first work done by the CCC enrollees included the following: clearing a large tract of land to make it suitable for campsites; constructing and grading a 900-foot road to the CCC camp from the National Pike; clearing and grading roads through the forest to improve access to fight forest fires and provide a way to public recreation areas; cutting brush and constructing fire-lines, including dividing the forest into thirty- to sixty-acre blocks to help contain forest fires to smaller areas; constructing a 112-foot-long bridge over White Sulphur Run on Fifteen Mile Creek Road using local timber; and opening up stone quarries, using the shale to build roads and improve the CCC campgrounds.

Joe Davis was responsible for training all enrollees in western Maryland in firefighting in both spring and fall. He introduced the "one-lick method" of fighting forest fires at Green Ridge State Forest. He said:

> *I came out with fire control notes. I was quite intrigued by the "one-lick method." So I began to experiment. I would have*

*groups of 25 enrollees. I would meet with them in the morning and talk fire theory. Then I would talk about the theory of fire fighting. Then in the afternoon, I would ring the fire-bell, they get in the stake-body truck, and we would fight an imaginary forest fire. Then I began to experiment with the one-lick method…I have the crew lined up…the machete man at the head, the brush men behind with their brush hooks, and then two men carrying the saw and carrying the brush away, and then maybe eight men raking, each man taking one swath and moving on and the other men coming through. Then behind them would be the fire tanks on their back and make sure the fire was going the right direction. Buckingham called me in one day and said, "Davis! What's this one-lock method you're using firefighting?" I replied, "Mr. Buckingham, I'm not using it, I'm just experimenting with it." He said, "Well don't use it!" I said, Well, O.K. Mr. Buckingham. A year later there was a Fire Wardens meeting at Green Ridge, it was around 1938–1939. There were 150 fire wardens there from Pennsylvania and Maryland. Buckingham calls me up and said, "Davis, would you mind bring up that one-lick method crew? We'd like to see that demonstrated." I said, "Yes, Sir!" So I had my crew of 15 men. Buckingham got an old firefighter from Swallow falls, Cecil Ramsey. He had been with the forest service fire-fighting for many years. So Buckingham outlined the fire boundaries. So we began to go. And my 15 men worked the one-lick method like a machine while Ramsey's crew was doing the old method of leap-frogging and so on. I think it was then that Buckingham accepted the fact that the one-lick method was a good method.*

The 1936 annual report of the Third Corps noted that "white enrollees" were the first to occupy the camp "prior to the coming of the colored enrollees on June 8, 1936." At this time, officials reorganized the Green Ridge CCC Camp S-53, 324 to S-53 MD-335-C (for "colored"). A pictorial CCC yearbook states that there were about one hundred enrollees. By this time, the camp had evolved to include a reading room, baseball and basketball teams, a dispensary, an education building, a vocational shop, a retreat and a

YOURS IN TRUST

WE MUST
PROTECT IT
FROM FIRE

"IF WE WOULD HAVE FORESTS — WE MUST PREVENT FIRES"

MARYLAND STATE DEPARTMENT OF FORESTRY

1939 Maryland Forest Service Forest Fire Prevention poster. *Courtesy of American Forests, 734 15th Street NW, suite 800, Washington, D.C. 20005. www.americanforests.org.*

camp singing quartet. Much of the educational activities took place in Cumberland, including classes in such courses as woodworking, photography, motion picture projection and soldering.

Officials established the Town Hill CCC Camp, also known as the Paw Paw CCC Camp, on June 14, 1933. The Paw Paw CCC Camp was located near the intersection of present-day Mertens Avenue and Oldtown Road. That camp's yearbook stated that one of the camp's most outstanding achievements up to 1936 was "placing crushed stone" on Oldtown Road from the Mertens Avenue intersection to Route 51, "a distance of five miles."

The last of the three CCC camps was established at GRSF on August 25, 1933, when officials established the Little Orleans CCC Camp (S-61) following the closing of the former Camp Ritchie CCC

Camp (S-55) near Frederick, Maryland. This camp was located off the Orleans Road and Mountain Road intersection, near present-day Interstate 68, exit 68, "about two miles from the Potomac River, up where Maryland manages to squeeze in between Pennsylvania and West Virginia."

The enrollees who were transferred from Camp Ritchie were accustomed to "palatial comforts" such as running water, an amenity lacking at their new camp at Little Orleans. Officials arranged transporting the needed water from the Green Ridge Camp ten miles to the Little Orleans camp. When the new enrollees arrived to

Adna "Pete" Bond, namesake of Bond's Landing at Green Ridge State Forest, in Maryland Forest Service uniform, circa 1942. *Courtesy Jenny Bond.*

a partial clearing in the forest, one of the first activities they observed at camp was digging a well. They finally hit water at a depth of about four hundred feet. Even though they were lacking some camp luxuries at the start, the enrollees worked on conservation-related projects involving constructing firebreaks, building roads and clearing and planting an eighteen-acre red pine plantation.

President Roosevelt said of the CCC program right from the beginning that "every boy should have the opportunity to work for six months in the woods. This [the CCC] was the happiest of the New Deal programs. For it simultaneously rehabilitated the land and the men." From Roosevelt's statement, the purpose of the CCC was not only to rehabilitate the environment but also to restore the spirits of young men. The *Forest Warden Newsletter* indicates that this is exactly what occurred at Green Ridge: "The men have certainly improved very much physically, and it is for certain that at least a part of the reason for their being in Camp has been accomplished. They will go back to Baltimore at the end of their term of enrollment better in many ways, and more able and self-reliant."

President Roosevelt, in a speech given over the radio on the third anniversary of the Civilian Conservation Corps, said that the CCC merited "the admiration of the entire Country." He praised the young men for the way they approached their daily work with a "fine spirit...winning the respect of the communities where their camps were located and through their spirit and industry...[they] demonstrated that young men can be put to work in our forests, parks, and fields on projects which benefit both the nation's youth and conservation generally."

Fred W. Besley said that the thousands of young men who served in Maryland's CCC camps advanced forest and park development by twenty-five years. The legacy of their work still can be seen today throughout Maryland's state forests and parks. Millions of forest and park visitors still benefit from the roads they built, the pavilions, bridges and stone culverts they constructed and the millions of trees they planted.

# Seventy Years of Camping and Hunting at Green Ridge State Forest

*The mountains are calling and I must go.*
*—John Muir*

For over seventy years, five generations of O'Neals have camped and hunted at Green Ridge State Forest. I recently stopped by the family's campsite on Howard Road to visit with them. On this particular day, brothers Bob and Jim O'Neal and their cousin Ron O'Neal were present.

"Your family's been camping and hunting in Green Ridge since 1936. What keeps bringing them back?" I asked.

Bob answered simply, "We are returning to the place of our youth. Every hollow and ridge holds a memory for us." Then he continued: "Our grandfather, Clarence Alonzo O'Neal, started it all. He was from Mount Savage. Rabbit hunting brought him to Green Ridge State Forest." I mentioned that in 1936 remnants of the famed Mertenses' apple orchard still existed. The orchard once provided great rabbit habitat. Today, hunters still kick up corrugated wire tree protectors from under the leaves on the forest floor where apple trees once grew.

Over the years, the O'Neals have camped and hunted primarily at three different locations within the Fifteen Mile Creek watershed at Green Ridge. They started with tent camping off M.V. Smith Road near Catpoint Road; then, in the 1940s, they converted a mule shed into a hunting cabin on the Shircliff property, a private tract in the forest. When their lease expired about 1969, the O'Neals returned to tent camping, mainly on Dug Hill Road and Howard Road.

"As kids we knew we were about to take 'the mountain trip' to Green Ridge when our grandfather announced it was time to go to camp," Jim said. The O'Neals explained to me that when school let out, they spent the first two weeks of their summer vacation at Green Ridge. "We were dropped off at our campsite and Grandfather O'Neal looked after us," Jim said. "Sometimes you'd find us wading in the Fifteen Mile Creek swimming hole; other times you'd find us fishing."

Grandfather O'Neal instilled respect for gun safety in his grandchildren. "We were told to break down a gun as soon as we walked out of the woods and unload immediately," Ron explained. "Grandfather would tell us: Don't point a gun at anything you don't expect to kill, don't shoot anything you don't expect to eat and know what you're shooting at and what's behind it."

Jim continued, "We started hunting with supervision as young teenagers—it was a rite of passage. We did shoot a few groundhogs at the age of nine, but we had to eat them."

The O'Neals have never hunted deer, choosing to stick with squirrel, grouse and turkey. I asked them how conditions have changed at Green Ridge over the years. According to Ron, grouse were more plentiful in the 1950s and '60s. In the' 50s, grouse habitat was better in much of the forest because it was in an earlier stage of development, providing ideal ruffed grouse habitat.

"Green Ridge State Forest was packed with hunters back then. You coughed to let others know you were around," Bob remembered. "We are proud that since 1936, no one in our family has received a citation for a hunting violation."

I asked the O'Neals if they ever observed a rare squirrel migration at Green Ridge. My research showed that the last great squirrel migration occurred in 1968 in the eastern United States. Bob recalled that as a teenager he observed what might have been such a migration. He remembered seeing large numbers of squirrels passing at one time through the forest and the older men saying that the animals were following the feed and moving on to another area.

I shared with them a naturalist's account recorded in 1811 of a vast migration observed in the Ohio Valley: "A countless multitude of squirrels, obeying some great and universal impulse which none can know but the Spirit that gave them being, left their reckless and gamboling life, and the ancient places of retreat in the north, were seen pressing forward by tens of thousands...to the South."

We paused to contemplate the wonder of it all. Later, as I made to leave, the O'Neals thanked me for visiting their campsite and taking an interest in them. However, I thought I should be thanking them, for it is through the significant contributions of dedicated outdoorsmen like the O'Neals that Maryland's forest health, public land acquisition and restoration of wildlife habitat efforts have been possible.

Aldo Leopold, Teddy Roosevelt and Gifford Pinchot are names that most people recognize as great leaders of the North American conservation movement; however, most people probably don't realize that they were also hunters. It has been said that hunters were the first conservationists. Hunters, anglers and outdoor enthusiasts have a close, personal attachment to the forests, fields and streams that support wildlife and fish habitat. Early in the twentieth century, hunters including Leopold, Roosevelt and Pinchot recognized that certain activities such as unregulated hunting, large-scale land clearing, wildfires and soil erosion were having dramatic impacts not only on our wildlife but on their habitats and forest health as well. Along with a growing movement of like-minded individuals, they saw the need for stricter laws and regulations, for government to manage and protect both wildlife and lands and for sustainable funding to carry out this mission. The abundant natural resources that we enjoy today and the public lands that help ensure access to them are a testament to their efforts.

In 1936, the O'Neal family began camping and hunting at Green Ridge State Forest; one year later, one of the most important pieces of federal legislation regarding conservation was passed. Known as the Pittman-Robertson or Federal Aid in Wildlife Restoration Act, it directs that 11 percent of the purchase price for firearms, ammunition and archery equipment go to the federal government and then to state natural resource agencies for wildlife conservation. As a result, hunters have contributed over $2 billion annually to national forest and wildlife conservation efforts since 1937.

Sportsmen have also contributed an estimated $185 million per year to forest and wildlife conservation through the purchase of hunting and trapping licenses or tags. Over the last century, it is estimated that hunters such as the O'Neal family have contributed over $5.5 billion toward forest and wildlife conservation.

Hunters continue to make significant contributions to Maryland's economy. According to a 2001 national wildlife survey, the estimated annual economic impact of deer, squirrel, turkey and grouse hunting statewide was about $301 million. In honor of Maryland Forestry and Parks' centennial year, we pay tribute to hunters like the O'Neal family for their considerable contributions to forest and wildlife conservation in Maryland.

# AN OASIS IN A MAD WORLD

*The Green Ridge German Prison of War camp was an "an island of peace in a crazy world...an island of human kindness in an ocean of hopelessness."*
—*Francis Boehm, Green Ridge German prisoner of war*

With the official closing of the CCC camps in 1942, little conservation work occurred for several years afterward. Most of the eligible men able to do this kind of work were serving in the armed forces during World War II. The next wave of available conservation workforce came from an unanticipated labor pool: captured German prisoners of war (POWs).

Americans generally are more familiar with the history of Japanese-American interment camps during World War II. However, America established five hundred POW camps between 1942 and 1946 that housed more than 400,000 prisoners from Germany, Italy and Austria. About 150,000 of these prisoners came from the German Afrika Korps captured in Africa in 1943. They were the first POW camps established on American soil since the Civil War. Officials also established POW camps in every state except Nevada, North Dakota and Vermont. Most POW camps were located in the warmer states to avoid heating expenses.

German prisoners were sent to the United States because England, where the prisoners were originally intended to be kept, was so overwhelmed with the number of POWs that officials requested the

United States' help with providing prisons. The Geneva Convention required feeding the prisoners properly. The large amount of food shipped from the States to feed the POWs overseas left a shortage of food to feed to American soldiers. Therefore, it made more sense logistically to send the POWs to America for housing. This also strengthened negotiations with enemies for better treatment and release of captured American soldiers.

In 1944, abandoned CCC camps across the country, which could accommodate up to two hundred men, were prime locations for POW camps. At Green Ridge, the former CCC camp along Fifteen Mile Creek became a German POW camp.

There was a labor shortage with the young men of the United States overseas serving in the military during the war with Germany and Japan. Due to the scarcity of available workers, apple growers in western Maryland requested help from the War Manpower Commission, stating that they needed people to work in the orchards. There were few workers to pick fruit or tend to the apple orchards in the area.

The Geneva Convention of 1929 dictated how POWs were to be treated. Their housing must be equal to the housing the United States provided its troops, with adequate heating and lighting. POWs could be assigned work assignments so long as they were not dangerous. Commissioned officers did not have to work, but if they were willing, they could volunteer to work. All POWs had to stay together no matter what their rank. Their work could not compete with American jobs. A labor shortage was the only justification that allowed POWs to work in the community, and their work had to help the war effort.

Most, if not all, of the POWs assigned to the Green Ridge POW Camp were a small part of the 150,000 captured in 1943 in North Africa. They were part of Rommel's German Afrika Korps.

At Green Ridge, two-and-a-half-ton army trucks transported POWs back and forth to work. The convoy included a few jeeps consisting of armed military guards. They generally worked no more than eight to twelve hours at a time. If a POW refused to work, he did not eat and was fed only bread and water until he started working again. They earned ten cents an hour, paid only in canteen checks and credit, never in cash. With these receipts, they could purchase items like toothpaste and toiletries.

When on work crews, the prisoners wore salvaged army fatigues with a large "P" on one arm and a "W" on the other arm to identify them as POWs. They could wear their German-issued uniforms only inside the camp compound but never on work details outside the camp.

At Green Ridge, most of the work was agricultural-oriented. Their primary task was to work on the apple orchards, the most desirable work from the perspective of a POW. Orchard work required pruning, cultivating and picking at harvest time. Perhaps the biggest benefit of orchard work for the POWs was that it required travel where they could temporarily leave the confines of camp and view the surrounding countryside, travel through towns and see the people of western Maryland. When traveling to sites to work outside the camp, POWs were loaded onto trucks fifty at a time with one guard and one driver. The guards counted the POWs getting on and off the truck. When back at camp, during recreational time, they played outside sports such as football and soccer.

POWs with disciplinary problems worked only in the nearby forest, cutting and loading pulpwood and timber, tasks that were considerably less desirable than working in the orchards. Bill Johnson, a Maryland Forest Service employee and future western Maryland district supervisor, remembered the strong character of the western Maryland forestry employee who had the ability to maintain order and control of any POW situation, commenting, "Our boys were tough as them."

Fortunately, few incidents requiring discipline occurred at the Green Ridge POW camp. Augie Diaz, a Cumberland resident and corporal with the 1322 Service Command at Fort Meade, was a prison guard at the camp. Recently, a *Cumberland Times News* reporter interviewed Diaz about his experiences as a POW guard: "It was a very good atmosphere there. There was good camaraderie between the guards and the prisoners and never any real episodes with anyone."

Diaz gave a talk to the students at the Green Ridge Youth Camp, located now at the former site of the former CCC and POW camp.

One of the youth recorded Mr. Diaz's presentation, writing that Mr. Diaz had served as a private first class prison guard from July 1944 until December 17, 1945. At the age of nineteen years, Diaz made twenty-eight dollars a month. The arrangement of the POW living quarters consisted of four separate barracks inside a twelve-foot-high wire fence. Inside the fence was a mess hall where the guards ate the same type and amount of food that the Germans ate. Diaz said that the Germans made the best lamb and potato cakes he has ever had.

Franz Boehm, a German POW assigned to the Flintstone camp, gave a personal account of his story to John Mash, former Green Ridge State Forest manager and author of the book *The Land of the Living*. Boehm arrived at the Green Ridge POW Camp in June 1944 and stayed there until December 1945. He was one of the last to leave before the POW camp permanently closed. Boehm said that his first feelings at the Green Ridge camp were "very romantic and one had the impression it was an old army fort of pioneer days"; however, "the reality was, we were not pioneers of old Europe, but war prisoners born of perverse men without God and faith who use and abuse men for their own madness."

Boehm remembered with affection the overall goodness of the Americans he encountered in Allegany County who displayed kindness, generosity and hospitality and a forgiving spirit to the German POWs at Green Ridge. Boehm noted that he received better treatment here as a POW than with his German peers in a combat environment. In Hitler's German army, food shortages, poor sanitary conditions, lack of proper clothing and the lack of freedom to practice one's faith were common. A preacher came on Sundays so that the POWs could attend a church service. Present was a translator to help them understand what was said.

Boehm celebrated two Christmas holidays at the Green Ridge POW Camp. He recalled that this was a very sad time for him, for had not seen his family in years. He remembers singing "Silent Night" in German with the others prisoners, the melody echoing in the wintry cold through the valley of Fifteen Mile Creek.

The POW camp at Green Ridge closed soon after the Christmas of 1945. Most of the German POWs returned to Europe; however, many of the POWs did not immediately return home after the war. First, as Mr. Diaz said, they had to serve "double-time" and went to places like France, England, Scotland and Belgium to help rebuild infrastructure destroyed by the Germans during the war.

Diaz was one of the guards assigned to transport the POWs back to France and from there to Germany. Mr. Diaz said that the prisoners at Green Ridge were loaded on the trucks and taken to a nearby train station and loaded on a troop train. From there, they traveled to New York for a two-week trip across the seas to France. The guards left them there and returned home to serve their remaining time. Unfortunate POWs sent to East Germany, under the control of Russia, disappeared forever.

Boehm and Diaz reconnected many years later and met. Diaz explained that he never really felt badly toward the prisoners. Like himself, POWs were recruited. They had not fought by choice.

Diaz still drives past the Green Ridge POW Camp occasionally, reliving in his mind his experiences as a prison guard. "It taught me to see the other side," Diaz said. "To be among them as a guard, I was able to observe them, see their habits, and their way of living. It was quite an experience that I will relish for the rest of my life."

# A Forest Under a
# Magnifying Glass

*Between every two pines is a doorway to a new world.*
—*John Muir*

The 1976 forest inventory differed from the earlier forest inventories of Allegany County, as this was the first forest inventory to focus entirely on Green Ridge State Forest. The purpose of the 1976 forest inventory was to determine the condition and composition of the forest at GRSF and develop improved methods to manage the state forest. The Maryland Forest Service, with stunning success, had implemented the forest conservation measures recommended by Sudworth and Besley. The recovery of the once devastated forest was evident from field information collected in the 1976 GRSF inventory.

John Mash, GRSF forest manager, oversaw the 1976 inventory. In his book, *The Land of the Living*, Mash observed that the forest composition had not greatly changed since pioneers settled the land more than 250 years ago.

In 1976, foresters established more than four hundred permanent continuous forest inventory (CFI) plots at GRSF on a two-thousand-by two-thousand-foot grid. Foresters and conservationists conducted several additional follow-up forest inventories in 1987 and the first decade of the twenty-first century, greatly relying on the CFI plots to collect their data.

The 1976 inventory, with GRSF at 36,438 acres, revealed that eight tree species made up 90 percent of the forest composition: white oak (24 percent); chestnut oak (18 percent); Virginia pine (13 percent); northern red oak (10 percent); black oak (10 percent); scarlet oak (9 percent); hickory—all species (5 percent); and white pine (1 percent). This was much the same forest composition that Sudworth and Besley had found in the early 1900s, although most of the forest in the early 1900s was in a seedling/sapling stage as compared to 1976, that show that the forest had matured.

The size class of the forest in 1976 was determined to be the following: seedling/sapling (2 percent); pole timber (59 percent); saw timber (51 percent); and non-forest areas made up primarily of managed open fields (18 percent).

Mash wrote this after evaluating the 1976 forest inventory data:

> *Ecologists often discuss the ever-changing forest, but within that context an underlying principle emerges—no matter what the catastrophe: fires, storms, insects, or man, Mother Nature heals all wounds, and puts things properly in her original perspective; there will always be forest. Not only will there be forest but it appears that the same species will make up succeeding forests. Diseases have removed some species from the woods, but one must keep in mind that the axe and saw never made a tree become extinct in America. They grow back.*

The official GRSF acreage recorded in the *1993 Green Ridge State Forest Management* Plan was 39,358 acres. The forest continued to improve in health, made up of the following size classes: seedling/sapling (10 percent); pole timber (47 percent); saw timber (38 percent); and non-forest (5 percent).

The forest inventory only hints at how remarkable, with its great tree diversity, are the forests of Green Ridge, a part of the larger Appalachian forests. Maurice Brooks wrote the following in his book, *The Appalachians*: "In this eastern broad leaved forest country, we take for granted [autumnal foliage color], not realizing that much of the world never sees such displays...[in fact] no other stands of deciduous forests in the midlatitude region(save, perhaps in China) are so rich and complex in woody species as [are the Appalachian forests]."

The fall foliage found in the forests of northeastern China and northeastern United States has a red hue that is lacking in the fall foliage of deciduous forests found in Europe, where it is more yellow or golden in color. Recent scientific discoveries indicate that the red colors evolved to protect the foliage from insects.

Scientific research published in *Live Science* explains that in "North America, as in East-Asia, north to south mountain chains enabled plant and animal migration to the south or north with the advance or retreat of the glaciers, according to the climatic fluctuations, and of course, with them migrated their insect enemies. In Europe, on the other hand, the mountains and its lateral branches reach from east to west."

Therefore, as the glaciers advanced in Europe, the trees could not migrate, as the mountains blocked their escape from the ice. Many tree species did not survive, along with the insects that depended on them. After the ice age, many of the European trees that did survive did not have to deal with the harmful insects now extinct. As a result, they did not produce red warning leaves, thus the reason the foliage on European deciduous trees is prominently more yellow or golden than the more colorful foliage of its American or Chinese counterparts that still must interact with a large array of harmful insects.

Likewise, spring in the Appalachian deciduous hardwoods is like no other place on earth, revealing a spectacular natural garden of beauty, especially when the trees and wildflowers reawaken after a long, cold winter. K. Abby Burk, MEd, and landscape manager at the Rocky Gap Lodge and Golf Resort in Allegany County, Maryland, said it beautifully: "Nowhere else does the monotone of the winter gray landscape fade into the rising currents of color that paint the mountainsides in living poetry as the mid-Appalachian in spring...nowhere in America does spring burst forth like it does in the northeastern United States." Indeed, it is something to behold in spring, the flowering of spring beauty, bloodroot and Virginia bluebell, along with serviceberry, redbud and dogwood, full of new life under the canopy of a cloudless blue sky.

The 1993 Green Ridge State Forest Management Plan reveals that GRSF was now a quite healthy mature forest, essentially a one-hundred-year-old even-aged forest. The bulk of GRSF, 90 percent, consisted of the following forest composition: mixed oak (58 percent); hardwood—white pine (6 percent); and mixed hard pine (4 percent).

The inventory revealed that the once devastated forest—burned, cleared, planted and converted into fruit orchards one hundred years earlier—had recovered! Foresters implemented at GRSF pioneering conservation practices recommended by Sudworth and Besley and other foresters who followed. The conservation practices worked with stunning success. Foresters implementing responsible scientific forest management practices proved the power of scientific forest conservation practices to restore and return the forest back to health.

# New "Old Growth" Found at Green Ridge

*This is the forest primeval.*
*—Henry Wadsworth Longfellow, American poet (1807–1882)*

P robably the most recognized old-growth stand in Maryland is the great hemlock forest at Swallow Falls State Park near Oakland, Maryland. However, this is not the only place old growth can be found in Maryland. Recently, several old oak stands, some approaching an age of four hundred years, were discovered at Green Ridge State Forest by the Department of Natural Resources (DNR) Old Growth Committee. "Old growth" is generally considered to be those forests that are relatively undisturbed, with a preponderance of old trees. The committee was formed to identifying old growth statewide on public and private lands. Members of the Old Growth Committee include natural resource professionals from a wide variety of backgrounds, including forestry, wildlife and the ecological sciences. A two-year grant from the U.S. Fish and Wildlife Service is funding this project.

By the early 1900s, most of the original eastern hardwood forests had been harvested. At Green Ridge, by 1918, the Mertens family of Cumberland, who owned the land at the time, had cut, burned and converted the forest into a large apple orchard. However, several areas unfavorable for growing fruit trees escaped the Mertenses' saw. Some of these areas were in coves that did not allow heavy cold air to circulate in spring, creating frost pockets. It was in these areas that the oldest stands of forest were found.

# New "Old Growth" Found at Green Ridge

Some values of old growth are associated with "beauty... spirituality...or a connection with the past." Words like "ancient" and "cathedral forests" are often found written on the same page with old growth. Old growth provides important wildlife habitat conditions not normally found in younger stands. For scientists, old growth is a valuable reference area for studying ecological functions and biological diversity.

The oldest tree found at GRSF was 366 years old. Trees like this are true sentinels of history. Just think of it—this tree, already over 200 years old in the 1700s, witnessed Thomas Cresap, Maryland's great frontiersman, roaming the land that is now Allegany County. Many years ago, passenger pigeons rested on its branches. The eastern elk, woodland bison, mountain lion and eastern timber wolf passed underneath its boughs.

Scientists can determine what the climate was like during colonial times through tree ring analysis. Based on their research findings, it appears that the 1500s experienced a "mega-drought." Scientists determined that the worst drought year in eight hundred years occurred in 1587. During that time, nearby, to the southeast at Roanoke Island, North Carolina, Sir Walter Raleigh established the first English colony in America. In 1587, Raleigh's colony mysteriously disappeared. Some historians now think that the megadrought had something to do with the mystery surrounding "the lost colony." This is just one example of how the study of tree rings can add to the interpretation of historical events.

Presently, the search for old growth has expanded from GRSF to Savage River State Forest and Potomac-Garrett State Forest. So far, fourteen old-growth stands, totaling approximately 187 acres, have been found at GRSF.

# MARYLAND'S BEST-KEPT SECRET

*John Denver could have been sitting here when he wrote the song "Country Roads," with the verse, "Almost heaven, West Virginia."*
*—Tom Darden, official photographer for five Maryland governors, speaking about scenic vistas looking into West Virginia from Green Ridge*

Describing a place as "one of Maryland's best-kept secrets" is overused, but in the case of Point Lookout at Green Ridge State Forest, it is true—it really is one of Maryland's best-kept secrets. Visitors to this location—not to be confused with the better-known southern Maryland state park of the same name—are always pleasantly surprised and taken aback when observing the spectacular view over the ancient Potomac River Valley in Maryland's Ridge and Valley Province as seen from Point Lookout.

Oldtown Road in GRSF, the oldest colonial road in eastern Allegany County, is the best route to travel on to reach Point Lookout. In 1758, during the French and Indian War, Colonel Thomas Cresap, Maryland's great pathfinder, blazed this road to improve and shorten the route between Fort Cumberland and Fort Frederick.

Some historians believe that the place name "Point Lookout" originated during the Civil War when Union troops stationed at neighboring Little Orleans protected the C&O Canal and B&O Railroad from Confederates intent on destroying the bridges and aqueducts along the Potomac River. Lookouts were established at Point Lookout and other nearby ridges, from which Union troops could observe the canal, railroad and Confederate movements throughout the valley.

Visitors to Point Lookout today can enjoy the same view that the Union troops had more than 140 years ago. From the overlook can be seen 243 acres of land that George Washington owned. The father of our nation traveled back and forth over what was then trackless wilderness more than sixteen times as a surveyor, landowner, military officer and later as the president of the United States. In the 1840s, the local people knew this serpentine section of the Potomac River as "General Washington's Horseshoe Bend."

In the early 1800s, the ownership log becomes a bit more complicated: partners Richard Caton of Catonsville lore and William Carroll of Rock Creek owned much of the land that is today Green Ridge State Forest, including Point Lookout. Caton was the son-in-law of Charles Carroll of Carrollton, a signer of the Declaration of Independence, while (William) Carroll was the grandson of Daniel Carroll, a framer of the U.S. Constitution. Located near Point Lookout off Carroll Road are the ruins of Carroll Chimney, built in 1836 as one of the nation's early steam-powered sawmills. It is the only surviving structure remaining from the period, as the duo's business ventures into iron ore and timber cutting eventually proved unsuccessful.

In addition to George Washington, another United States president, Abraham Lincoln, is associated with the view at Point Lookout. From Point Lookout one can see in the far distance the tracks of the former B&O Railroad, where Abraham Lincoln passed through in 1847. John E. Clark wrote in an article, "The Transportation Revolution in Illinois," that in November 1847, Abraham Lincoln and his wife Mary made a twelve-day, 1,500-mile journey from Illinois to Washington to take his seat as a newly elected congressman: "They took the stagecoach over bone-jarring roads from Springfield to St. Louis. A steamboat then carried them up the winding Ohio River to ride to Wheeling, then part of Virginia. They changed in Wheeling to another stagecoach. It took them over the National Road to Cumberland, Maryland, where the Baltimore and Ohio Railroad whisked them on the final leg to Washington."

In 1863, Lincoln permitted this section of former Virginia to secede from the Confederacy. On June 20, 1863, West Virginia became part of the Union, the only state to secede from the Confederacy. In this section of former Virginia, many people were sympathetic to the Union cause in the Civil War. Admission to the Union was under the condition that the new state of West Virginia would write in its constitution language that gradually abolished slavery.

In the late 1800s, the Mertens family of Cumberland acquired Point Lookout and the Green Ridge property from the Carroll family. The Mertenses cut, burned and converted the forest into a fruit orchard, which they promoted as "the largest apple orchard in the universe." They then subdivided the land into more than three thousand ten-acre lots and sold each lot to individual owners all across the country. To impress potential buyers, the Mertenses' first stop on their orchard tour was Point Lookout. The family proclaimed the overlook as the most beautiful spot on the East Coast: "I can say that from Point Lookout is the most beautiful view my eyes have ever fallen upon. It excels Pike's Peak or any other...There is no view equal to it." In 1918, the Mertenses fell into bankruptcy and abandoned the orchard and Point Lookout.

The State of Maryland purchased a little over 2,000 acres of land in 1931 in eastern Allegany County, establishing Green Ridge State Forest. More than seventy-nine years later, the state forest has grown to 43,560 acres that include Point Lookout. The Department of Natural Resources established the area around Point Lookout as wildland, thus protecting the view on the Maryland side.

Point Lookout, once part of the "largest apple orchard in the universe," is now part of Green Ridge State Forest, Maryland's largest contiguous block of forestland within the Chesapeake Bay watershed. As one of the Mertens family's publicity men described the area roughly one hundred years ago, "Tourists have traveled thousands of miles and then pronounced it the most beautiful view in the United States, yet there are few people in Maryland who know the place exists."

For those who love to find beauty and the sublime in the great outdoors, when at GRSF, one must visit this place. Photographs cannot possibly capture all the spectacular scenery or the magnificence concentrated here. Still not well known, it is just one more reason Point Lookout is one of Maryland's best-kept secrets. Consider a visit and discover the grand wonders there for yourself.

# George Slept Here Too

*At 15 Miles Creek…recrossed the Potomack…to a tract of mine on the Virginia Side…having reviewed this land I again crossed the river and getting into the Waggon Road* [present Oldtown Road in Green Ridge State Forest] *pursued my journey to…old Town…lodged at Colo. Cresaps.*
—*George Washington, September 8, 1784*

Thomas Jefferson's father, Peter Jefferson, a pioneering surveyor, was one of the first to put Fifteen Mile Creek on the map. In 1751, Peter Jefferson, with his partner Joshua Fry, produced the map known today by historians as the Fry/Jefferson map. The designation of Fifteen Mile Creek on this map makes it one of Allegany County's oldest place names.

One person very familiar with the Fifteen Mile Creek landmark was George Washington. His journal shows more than sixteen visits to this region of the country throughout his life as a surveyor, military officer, landowner and president of the United States. Washington's quote at the beginning of this article is taken from his journal, written not long after the American Revolutionary War.

His success as a general of the colonial military forces made Washington one of the most famous people in the world, yet we find him traveling alone the day of September 8, 1784, meeting a party on the banks of Fifteen Mile Creek and the Potomac River. On this particular trip, Washington was inspecting his tracts of western land on the Virginia side of the Potomac River. Once he completed his inspection, Washington knew that he needed to travel only fifteen miles west by road from the junction of Fifteen Mile Creek and the Potomac River (thus its place name, Fifteen Mile Creek) to reach the frontier town of Oldtown, where he planned to stay the night with Colonel Thomas Cresap.

Fifteen Mile Creek's headwaters originate in Pennsylvania and travel 19.3 miles before emptying into the Potomac River. Buffered almost entirely by trees and state forests, Fifteen Mile Creek is one of Maryland's most pristine streams. A section of Fifteen Mile Creek is categorized by the Maryland Department of Environment as a Tier II Stream, indicating that its superb water-quality conditions are better than necessary to support fishing and swimming.

The plant harperella is listed as federally and state endangered, and it grows on Fifteen Mile Creek, one of only about twenty places on earth where it is known to exist. To survive, harperella requires fluctuating water levels during its life cycle. In the spring, harperella needs moderate intensive flooding to scour rock bars and crevices of competing vegetation. In summer, the opposite is needed; harperella then requires periodic low water flows in order to expose its flowers and set seed.

Working at Green Ridge State Forest for more than twenty-five years, I have observed Fifteen Mile Creek during times of floods and during times of drought. In spring, Fifteen Mile Creek can flash with high water flows; then several days later, it quickly reverts to a slow trickle of flowing water. In the summer, during long drought periods that last months, I have seen Fifteen Mile Creek look like a dusty, dry, rough cobblestone road where an occasional pothole is found filled with water. Stream tributaries that feed into Fifteen Mile Creek (Pine Lick, Piclic, Terrapin Run and Mudlick) do not provide much help, as they are also bone-dry in the summer months. Human-caused long-term disturbances to this seasonal pattern of flow can degrade harperella's habitat and fundamentally alter the nature of the streams that George Washington once knew. Bringing the gift of water to Maryland's driest region, the growing scarcity of pristine streams like Fifteen Mile Creek makes them that much more valuable.

A famous doctor once said that our top job is to be good ancestors. In a state that is developing quickly, hopefully we will make informed decisions that benefit future generations so they too have pristine streams like Fifteen Mile Creek to enjoy and cherish.

# GOING TO THE MOUNTAINS IS GOING HOME

*Over civilized people are beginning to find out that going to the mountains is going home.*
*—John Muir*

On November 17, 2007, the Great Eastern Trail (GET) Association held its first official meeting at Hungry Mother State Park in Marion, Virginia. Present were members from a coalition of volunteer trail clubs, including the Friends of Green Ridge State Forest, the Potomac Appalachian Trail Club, the Standing Stone Trail Association, the Mid-State Trail Association, the Pine Mountain Trail Conference and the Appalachian Trail Conservancy. Hiking club members approved the bylaws of the GET organization, including its mission and purpose statement: "To conceive, create, build, develop, and promote the Great Eastern Trail; and to educate the public in the use and appreciation of the GET and all aspects of it natural surroundings."

The GET organization favors hiking; however, local trail managers will have the ultimate say on whether it is used as a single or multiuse recreational trail. Regardless of its designation, here hikers can trek far into the interior of the forest, where they can realize the beauty of the earth.

The concept of the GET is an old idea originally conceived by Benton MacKaye, a forester, regional planner and founding father of the Appalachian Trail (AT)—the crown jewel of all hiking trails on the East Coast. MacKaye and Fred W. Besley, Maryland's first state forester, were professional associates, both having trained and worked under Gifford Pinchot, the father of American forestry. Besley and MacKaye also attended Pinchot's famous Baked Apple Club meetings and were lifelong members of the Society of American Foresters.

An original 1920s map still survives of the Appalachian Trail drawn by MacKaye. His primary vision was a trail that followed the crests of the ancient Appalachian Mountain chain extending from New England to the South. On MacKaye's map are drawn several lines extending west from the AT. In Maryland, this map shows a line going west from the AT following the corridor of the C&O Canal. This line abruptly shoots north through present-day GRSF and appears to connect to the present-day Mid-State Pennsylvania trail. This is the present alignment of the GET in Maryland. This map is evidence of MacKaye's extraordinary vision, as the C&O Canal National Historic Park and GRSF did not yet exist as public lands.

The majestic vision behind the GET is awe-inspiring. The GET will connect the Finger Lakes Trail in New York to the southern terminus of the Pinhoti Trail in Alabama. Planners are also exploring ways to connect the Alabama Pinhoti Trail to the Florida Trail. Just think of it—once the GET becomes a reality, a person, with a superabundance of endurance, can hike all the way from New York to the Florida Keys!

One major challenge for the organization is to connect several large gaps along the GET to make it one continuous, uninterrupted trail. Once completed, the GET will be more than 1,800 miles long and will link more than 10,000 miles of trails along its route from New York to Florida.

The quote introducing this article was used by Benton MacKaye in his book, *A New Exploration*, published in 1928. Here, MacKaye discussed his philosophy of regional planning and the need to preserve open space for the well-being of the growing population in the United States. MacKaye anticipated a growing population of city dwellers along the eastern seaboard and their need to escape from the city for outdoor recreation. He foresaw that the general public living in urbanized environments would need places like hiking trails to reconnect with nature. In MacKaye's mind, open spaces would provide havens to escape the hectic pace of city life, providing quiet places to restore and rejuvenate the body, mind, heart and spirit. I wonder if MacKaye and Besley foresaw the miles and miles of bumper-to-bumper westbound traffic on late Friday afternoons escaping Washington, D.C., and Baltimore and "going to the mountains" on I-70 and I-68.

Benton MacKaye's vision for the Appalachian Trail took sixteen years (1921–37) to come to fruition. Today, it is estimated that more than three million hikers use the AT each year. Hopefully, the vision for the GET will become a reality just as quickly, enticing millions more to "go to the mountains." Go to www.greateasterntrail.org to learn more about the GET.

The idea that Green Ridge State Forest would become part of the Great Eastern Trail system is quite exciting, considering its humble origins. Beginning about 1974, John Mash, forest manager at GRSF, began a monumental project to design, map and build a hiking trail that would extend from north to south the entire width of Maryland in eastern Allegany County, from the Pennsylvania/Maryland line to the Potomac River.

The construction of the hiking trail was made possible with the assistance of GRSF staff, Juvenile Services, the Youth Conservation Corps (YCC) and, later, the Young Adult Conservation Corps (YACC) and volunteer organizations. When the work was completed, workers had constructed a twenty-four-mile network of linear hiking trails at GRSF.

During this time, the staff built two attractive suspension bridges over Fifteen Mile Creek. A hiking trail brochure was printed showing connections to the C&O Canal National Historic Park trail. This gave a willing hiker the opportunity to hike a forty-three-mile loop, which usually takes about three days.

The hiking trail became very popular with outdoor enthusiasts. Besides routine trail maintenance, the staff made very few changes to the trail until 1996, when a flood destroyed the two suspension bridges. This event energized efforts to rehabilitate the hiking trail. Financial support provided from the Recreational Trail Grants program, administered by the State Highway Administration, made it possible to restore the trail between the years 1996 and 2006. During this time, along with routine maintenance of the trail, Maryland Forest and Park Service staff, with help from Juvenile Services, the Maryland Conservation Corps and many volunteers, built two new bridges over Fifteen Mile Creek and three Adirondack shelters and produced a professional brochure along with interpretive trail exhibits highlighting the "Leave No Trace" message.

Accolades began accumulating like a growing snowball gathering more snow as it rolls down a steep backwoods trail. In 2003, GRSF staff received a national award on Capitol Hill in Washington, D.C., from the Coalition for Recreational Trails for "outstanding use of recreational trails program funds in education and communication." The Green Ridge Hiking Trail was becoming more popular.

Meanwhile, at GRSF, hikers were requesting a shorter circuit trail loop than the forty-three-mile circuit offered at the time. On May 24, 2004, the Twin Oaks Trail, a four-mile circuit trail, was dedicated. This circuit trail was the first to be added to the GRSF trail system since John Mash implemented the original trail in 1974.

In June 2005, with support from the National Park Service, the Green Ridge State Forest hiking trail system was officially designated a National Recreation Trail, becoming part of the Potomac Heritage National Scenic Trail system.

In May 2006, at a meeting coordinated by the American Hiking Society in Blacksburg, Virginia, the GET organization became an official entity. At this meeting, Green Ridge State Forest National Recreation Trail was accepted as part of the Great Eastern Trail system.

The Great Eastern Trail takes hikers past ancient geologic features and waterfalls and to spectacular views from ridge tops. They will also see a rich diversity of trees, wildflowers and wildlife. It takes hikers deep into the interior of forests where, as the writer and naturalist Henry Beston wrote, they can "touch the earth...honor the earth and rest [their] spirit in her solitary places."

Along with social and environmental benefits, trails like the GET help the local economy by bringing more people into the area. Hikers will invest in nature tourism and support local merchants by purchasing food and lodging and attending local cultural events. Hopefully, trails like those found at GRSF and along the GET will help produce a growing corps of environmental stewards and outdoor recreationists who will continue to enthusiastically support both forest conservation and open space.

# THE INCREDIBLE FLIGHT OF
# THE BROAD-WINGED HAWK

*Doth the hawk fly by thy wisdom and stretch her wings to the south?*
 —*Job 39:26*

Since ancient times, people have pondered the mysteries of bird migration with wonder and awe. Where do birds fly to during the fall migration? How far south do they actually go? Where do they spend winter? In our own lifetime, thanks to science and advances in technology, we are obtaining answers to these questions. Some of the research to advance our understanding about bird migration took place recently at Green Ridge State Forest.

During the spring of 2000, scientists Jim Dayton and Jack Cibor, members of Earthspan, Inc., with assistance from the Department of Natural Resources, trapped an adult, female broad-winged hawk at GRSF. Once captured, the hawk was quickly fitted with a small backpack containing a high-tech Service Argos Platform Transmitter Terminal (PTT) and then released back into the wild. The PTT allowed the scientists to track the hawk's movements with satellites orbiting 1,368 miles above Earth. This was the first time scientists had ever used this method to track the migration of a broad-winged hawk. Scientists designated the GRSF hawk with the name "MD93." This distinguished it from three other hawks that scientists were tracking at the same time (one from Savage River State Forest, Maryland, and two from north-central Minnesota).

About September 10, 2000, MD93 began its fall migration southward from GRSF. It followed the Appalachian flyway over Louisiana and southeastern Texas and continued south, following the Gulf of Mexico shoreline. By late September, MD93 had passed over Veracruz, Mexico, continuing south on an inland course over Colombia through central South America. Finally, about December 15, after almost three months, covering about 62 miles per day and traveling a total 4,100 miles from GRSF, MD93 arrived in southern Peru, where it spent the winter.

MD93 was the only hawk returning to Maryland that still had its PTT attached when the time came for spring migration. This provided scientists the opportunity to track MD93's flight back to Maryland. About March 11, 2001, MD93 left Peru, flying north on the same pathway it had taken during the fall migration. Seventy-four days later, about May 21, after traveling an average of 65 miles a day and a total distance of 4,889 miles, MD93 returned to GRSF. Remarkably, it made a new nest within a mile from where its old nest had been the year before!

In 2003, Earthspan published its research in the *Wilson Bulletin* in an article titled "Migration Routes and Wintering Locations of Broad-Winged Hawks Tracked by Satellite Telemetry." Much of the information for this article came from this document. Scientists have learned in studies like this that birds often follow a narrow flight path when migrating. Information from studies like this will help scientists develop more effective conservation strategies along the flyways these birds follow.

Members of Earthspan also received credit for contributing their peregrine falcon migration research to the April 2004 edition of *National Geographic* magazine. This edition has an article about bird migration and includes a beautifully illustrated bird migration map. Interestingly, the map shows that the birds of prey migration route along the Appalachian flyway is similar to the one taken by the GRSF hawk.

The *National Geographic* article describes bird migration as "one of nature's most moving performances." The article informs us that birds, like the hawks at GRSF, have been following the same ancient flyway for at least fifteen thousand years, since the last ice age.

# The Incredible Flight of the Broad-winged Hawk

The GRSF hawk migrated more than 8,500 miles round trip. Earthspan's research reminds us how remarkable it is that these birds have the strength to migrate such long distances on "wings woven [only] of delicate feather and hollow bone."

Aristotle, one of the earliest naturalists on record to write about bird migration, perhaps had this in mind when he wrote, "In all things of nature there is something of the marvelous." If he were alive today, I think the old Greek philosopher himself would marvel at Earthspan's high-tech scientific research methods and the astonishing findings they made regarding the incredible flight of the hawk from Green Ridge State Forest.

# SCIENTISTS BRING GREEN RIDGE INTO THE SPACE AGE

*I believe that the natural resources of Allegany County are some of the best in the country.*
—*Janice Keene, owner of the Evergreen Heritage Center Foundation, Mount Savage, Maryland*

On November 21, 2000, at the beginning of the second millennium, a 7320 Delta Rocket launched deep into outer space from Vandenberg Air Force Base, California. This launch occurred exactly one hundred years after Sudworth's 1900 benchmark forest inventory in Allegany County, Maryland. The rocket launch initiated a new era of scientific forestry pioneering, involving modern space-age techniques and tools. Green Ridge State Forest served as one of the main areas of focus where this rocket-age, modern-day scientific trailblazing work took place.

The Delta Rocket carried on it the Earth Observing-1 (EO-1) satellite, NASA's first operational hyperspectral imager put into orbit. The technology of the Hyperion sensor on the EO-1 satellite gave it the ability to study earth's environment and ecology from afar in outer space. The goal of using this new satellite technology was to develop new techniques to improve the performance of future earth science missions and, in this case, improve forestry data analysis from satellite remote sensing equipment.

An article in the *Journal of Forestry* published in June 2000 notes that foresters pioneered remote sensing technology: "They were among the pioneers in research leading to advances in many of the methods and tools used in this field...mapping forest vegetation from aerial photographs was first attempted in the 1850's using a camera carried aloft on an air balloon." Foresters' early experience with aerial photographs taken from airplanes in the 1920s and 1930s made them especially valuable "as photo interpreters during World War II." By the 1950s, most forestry schools offered classes in air photo interpretation. Printers produced aerial photographs before the 1960s in black-and-white.

Green Ridge State Forest office still uses original black-and-white aerial photographs from 1937, stored there since Fred Besley's administration, to interpret the forest landscape. In fact, it was the 1937 aerial photographs that helped foresters recently locate more than two hundred acres of old growth at Green Ridge State Forest. By the 1960s, photographs taken from airplanes were produced with color photography that "permitted foresters to look at the terrain in four spectral colors...the blue, green and red bands of normal color film." The collection of 1983 color aerial photographs at Green Ridge were an important tool in developing the 1992 forest inventory and were especially helpful in locating and inventorying pine stands located on the forest.

In 2000, however, the technological advance of remote sensing was about to accelerate as fast as the Delta Rocket that transported the EO-1 carrying the Hyperion in circular orbit 438 miles above the earth. Hyperion was the world's only hyperspectral satellite sensor in space in 2000. It was an extraordinary instrument with the ability to measure reflected light from Earth's surface in 220 spectral bands, an immense improvement of only the 4-color bands available since the 1960s.

Scientists from NASA and the U.S. Geological Survey collaborated on this project to determine whether the Hyperion imager in space would be more efficient and effective in evaluating forest vegetation composition and its condition from space than the more earthbound methods of collecting forest data.

Because of GRSF's proximity to the Appalachian Laboratory, an environmental research facility of the University of Maryland Center for Environmental Science located at Frostburg, Maryland,

GRSF became one of the first places on earth to test the space-age forest cover imaging technology that NASA developed. Dr. Philip A. Townsend from the Appalachian Laboratory wrote and eventually was awarded the grant by NASA. Dr. Townsend and his assistant, Jane R. Foster, represent a new generation of forestry pioneers made from the same trailblazing cloth of Sudworth and Besley. They oversaw and conducted their work at Green Ridge State Forest, testing the forest inventory value of hyperspectral imagery technology for the first time.

The primary purpose of this forest inventory study was to test the Hyperion imagery and data from space to map forest composition to determine its ability to accurately predict forest community types on earth from space. Townsend and Foster worked with other scientists on a national validation team located in other parts of the country. Together they tested, assessed and validated the data to determine its practical use for forestry and other applications. The Appalachian Laboratory interpreted that data and communicated the information back to NASA. At the same time at NASA, scientists also identified and recommended corrections to help make this technology available to a future wider audience.

The technology works something like this: each tree species, depending on its age and condition, reflects light uniquely across visible (colors) and invisible (infrared) wavelengths. By linking these unique reflectance signatures to the tree species found in the forest, it becomes possible to inventory the forest from space. Jane Foster analyzed the images produced by Hyperion with help from forest plot data collected by Maryland Department of Natural Resources personnel. Jane Foster also further checked the space-derived data at GRSF in the field with additional on-the-ground "field-truthing." Foster stated that while she roamed the woods of Green Ridge, she was surprised to see the abundance of black oak growing there. That black oak is so abundant says much for the ability of its bark to protect it from wildfires that overran the forest one hundred years ago in the late 1800s and early 1900s, while other more susceptible trees fell out of the forest population.

Instead of hand-drawn maps, the way foresters worked in prior forest inventories, new-age foresters used computer digital software

Jane R. Foster, scientist for Appalachian Laboratory, University of Maryland Center for Environmental Science, Frostburg, Maryland. *Courtesy Jane R. Foster.*

mapping programs to make forest inventory maps. Foresters also used global positioning systems (GPS) to communicate with satellites hundreds of miles in space that identified exact locations of forest plots and improved the accuracy of the maps.

Foster identified twenty-eight different forest community types in this inventory and broke this analysis down further into thirteen common forest types: These common types were identified by dominant overstory species as: red oak; red oak–oak–hickory mix; Virginia pine–deciduous mix; chestnut oak–scarlet oak; scarlet oak mix; pitch pine; black oak mix; white oak; chestnut oak–oak mix; chestnut oak; white pine; white pine–oak mix; and eastern hemlock.

The forest inventories that have occurred in Allegany County and GRSF bring to light not only that the first scientific forest inventory occurred in Allegany County, but also that throughout the last century, Allegany County and GRSF hosted some of the leading pioneers in forest conservation. The inventories also tell of the rich diversity and wonderland that the forest is. The foresters' recommendations of the past were followed over the years: timber harvests were regulated, wildfires were controlled and livestock were kept out of the woodlands, resulting in the restoration of devastated forested landscape back to health, benefiting society with gifts of clean water, timber, wildlife and places for outdoor recreation. As Janice Keene noted at the heading of this chapter, the forests, once devastated, are now restored to a place of great value.

In 1912, Besley said that the forests were the most important feature of Allegany County. One hundred years later, forests are still what distinguish Allegany County from all Maryland counties east of it. That Allegany County contains GRSF, one of the state's natural crown jewels, makes it that much more special.

Offutt Johnson, a historian of Maryland's state forests and parks, once said, "Maryland is like a beautiful woman who is adorned with a necklace of emeralds of different sizes. The emeralds on the necklace represent all of Maryland's state forests and parks. All are valued jewels; GRSF is one of the larger jewels on the necklace, being the largest contiguous block of forests within the Chesapeake Bay Watershed in Maryland."

William Penn Mott, director of the National Park Service under Ronald Reagan, added to the concept of public lands as natural jewels: "Indeed, crown jewels are spectacular, but all parks and forests, regardless of size and features are special places. Indeed they are all shining jewels on the crown." With the help of forest inventories and acting on the foresters' recommendations derived from the inventories, Green Ridge State Forest today is one of the shining jewels within the system of state forests and state parks in Maryland.

# FORESTRY AND THE PIONEERING WORK ETHIC

*Every man in his heart revolts at civilization and will revert back to* [nature] *if given half a chance...We don't live long enough to find out what life is all about, but we know what civilization is—it is a mere veneer that keeps on getting thicker, but never too thick to pierce... It will be 15,000 years I think, before man will reach such a high point of civilization where he cannot and will not want to go back to* [reconnect with nature].
—*Thomas Edison while camping in western Maryland, July 1921*

Based on Thomas Edison's quote at the head of this chapter, we are going to need places like Green Ridge State Forest to reconnect with nature for at least the next fifteen thousand years. Fortunately, because of the leadership skills of our pioneering conservationists of the past, places like Green Ridge State Forest exist.

James P. Owens identified the ten principles of "cowboy ethics" from two books he wrote, one called *Cowboy Values: Recapturing What America Once Stood For.* The cowboy ethics are the following: to "live courageously, take pride in their work, finish what they start, do what's necessary, be tough but fair, keep promises, ride for the brand, talk less and say more, remember some things aren't for sale, and know where to draw the line." This was also the code of our frontier and forestry pioneers mentioned in this book. These are the same principles that made them successful and will make present-day pioneers successful.

The legendary western American cowboys are descendants of the frontier pioneers like Boone, Crockett and Cresap, who roamed the lands east of the Mississippi River. The cowboys inherited the principles they applied out West from their pioneering ancestors in the East.

An old photograph taken of George B. Sudworth in 1901, one year after his Allegany County forest inventory, conducting fieldwork in the western part of the United States shows him sitting on a mule dressed like a cowboy with hat, holster and pistols, with a rifle mounted on the saddle. Indeed, back then, foresters were much like cowboys. Pinchot required that his foresters know how to ride a horse and/or mule, camp and take care of themselves for long periods in the great outdoors. However, there was one big difference between cowboys and foresters back then: cowboys pushed cattle on the open plains, and their primary interest was to get cattle to market; foresters pushed cattle out of the woodlands onto the open plains, and their primary interest was to protect forest regeneration and restore the forest landscape.

It is a fact the Fred W. Besley passed the cowboy ethic not only through his employees but down through his family as well. His granddaughter, Mary Rotz, annually holds the Cowboy Christmas show at the Antietam's Recreation near Hagerstown, Maryland. Among the performers is Besley's great-grandson, Andy Rotz, a master handler of the rope, bullwhip and knives. He also holds the Guinness Record for doing 11,123 consecutive Texas skips, where he spins a horizontal rope loop and jumps back and forth. The incredible feat took him three hours and ten minutes when he broke the record in Las Vegas.

Foresters conserving forest also conserved wildlife habitat, providing havens of refuge for wildlife. One of the great conservation success stories of the last century involved bringing back many animals once on the verge of disappearing from the Green Ridge State Forest landscape, such as white-tail deer, wild turkey, black bear and ravens, to name a few. Theodore Roosevelt would shout out with a loud enthusiastic "Bully!" if he could see today the success of one of his broad goals realized since he first sent foresters like Fred W. Besley out into the field from Gifford Pinchot's living room in

Washington to restore the devastated forested landscape more than a century ago. It is as true today as when Roosevelt said it in 1902: "We have taken forward steps in learning that wild beasts and birds are by right not the property merely of people alive to-day, but the property of unborn generations, whose belongings we have no right to squander."

Pioneers are leaders. They are trailblazers and pathfinders, leading the way to new places of exploration and discovery. James M. Kouzes and Barry Z. Posner are credited with the following quote: "Leaders take us to new places we've never been before. But there are no freeways to the future, no paved highways to unknown, unexplored destinations. There's only wilderness."

The pioneers mentioned in this book have led us to the present, and as the first one hundred years of forest conservation conclude in Maryland, many gifts along the trail have been given by them to us, one being Green Ridge State Forest. A new generation of pioneers has stepped up, and they are leading us now into the second century of forest conservation. As other pioneers have done in the past, they will stand on the shoulders of others to see further and lead us successfully onward to the unexplored wilderness and promised land of tomorrow, all along the way preserving and conserving our precious natural resources for future generations.

# BIBLIOGRAPHY

*One touch of nature makes the whole world kin.*
—*William Shakespeare*

Bailey, Kenneth P. *Thomas Cresap: Maryland's Frontiersman.* Boston: Christopher Publishing House, 1944.

Bailey, Robert F., III. *Maryland's Forests and Parks: A Century of Progress.* Charleston, SC: Arcadia Publishing, 2006.

Besley, F.W. *The Forests of Allegany County.* County Forest Inventory Report. Baltimore: Maryland State Board of Forestry, 1912.

———. *The Forests of Maryland.* Annapolis, MD: Press of the Advertiser-Republican, December 1916.

Brauer, Norman. *There to Breathe the Beauty: The Camping Trips of Henry Ford, Thomas Edision, Harvey Firestone, John Burroughs.* Dalton, PA: Norman Brauer Publications, 1995.

Brinkley, Douglas. *The Wilderness Warrior: Theodore Roosevelt and the Crusade for America.* New York: HarperCollins, 2009.

Brooks, Maurice. *The Appalachians: The Naturalist's America.* Boston: Houghton Mifflin Company, 1965.

Civilian Conservation Corps. *District Number 2, Third Corps Area. Official Annual.* Yearbook. Baton Rouge, LA: District Advertising Company, 1936.

———. *Editorial Review, Northern District, Third Corps Area, Company 335-C, Camp Green Ridge.* Yearbook. Atlanta, GA: Army-Navy Publishers, Inc., 1933.

Cresap, Joseph Ord, and Bernarr Cresap. *The History of the Cresaps.* Gallatin, TN: Cresap Society, 1987.

F. Mertens and Sons. "Announcement." Promotional pamphlet. Baltimore, MD: Munder-Thomsen, 1910.

———. "The Commercial Apple Industry in the United States." Promotional pamphlet. Baltimore, MD: Munder-Thomsen Press, 1910.

———. "Commercial Department of Apple Culture." Promotional pamphlet. Baltimore, MD: Munder-Thomsen Press, 1910.

———. "Report on Green Ridge Valley." Promotional pamphlet. Baltimore, MD: Munder-Thomsen Press, 1910.

Maryland Department of Natural Resources. *Forestry Towers and Forestry Tower Properties.* Administrative report. Annapolis: Maryland Department of Natural Resources, Public Lands Policy and Planning, December 2006.

———. *Green Ridge State Forest Ten Year Resource Management Plan, Vol. I and Vol. II.* Offical document. Annapolis: Maryland Department of Natural Resources, 1993.

———. "Maryland's Conservation History." March 28, 2010. http://dnr.maryland.gov/mdconservationhistory.

Mash, John. *The Land of the Living: The Story of Maryland's Green Ridge Forest.* Cumberland, MD: Commercial Press, 1996.

Platt, Rutherford. *The Great American Forest.* Englewood Cliffs, NJ: Prentice-Hall, Inc., 1965.

State Department of Forestry. *Maryland Forest Warden Newsletter.* Monthly forest warden report. Baltimore, MD: State Department of Forestry, 1932–33.

Sudworth, George B. *The Forests of Allegany County: Maryland Geological Survey.* Baltimore, MD: Johns Hopkins Press, 1900.

United States Government. *Second Report of the Director of Emergency Conservation Work.* Annual report. Washington, D.C.: Government Printing Office, April 5, 1933, to September 30, 1933; and October 1, 1933, to March 31, 1934.

Van Deman, Professor H.E. *Apples in Western Maryland.* Promotional pamphlet. Baltimore, MD: Munder-Thomsen Press, 1910.

# ABOUT THE AUTHOR

C hamp Zumbrun recently retired from the Maryland Department of Natural Resources, where he worked more than thirty years, with most of that time spent as forest manager at Green Ridge State Forest. He and his wife, Cindy, reside in LaVale, Maryland, where they raised two sons, Jeffrey and Ryan. Zumbrun is a licensed professional forester who obtained his BS in forest resource management at West Virginia University and an MS in management at Frostburg State University in Maryland.

This photograph was taken during a scientific DNR Wildlife Administration black bear inventory at Green Ridge State Forest.

He has written a column for the *Cumberland Times-News* and articles for the Department of Natural Resources, the Forest History Society and local historical organizations. When time permits, he enjoys playing guitar, composing and recording songs.

Please visit us at
www.historypress.net

www.ingramcontent.com/pod-product-compliance
Lightning Source LLC
Chambersburg PA
CBHW070757300326
41914CB00053B/697